International Conference on Progress in Mechanical and Aerospace Engineering (PMAE 2018)

MATEC Web of Conferences Volume 257 (2019)

Hong Kong, China
10 - 12 August 2018

Editors:

Wei-Hsin Liao
John Mo

ISBN: 978-1-5108-7985-0

Printed from e-media with permission by:

Curran Associates, Inc.
57 Morehouse Lane
Red Hook, NY 12571

Some format issues inherent in the e-media version may also appear in this print version.

For additional information, please contact EDP Sciences – Web of Conferences
at the address below.

EDP Sciences – Web of Conferences
17, Avenue du Hoggar
Parc d'Activité de Courtabœuf
BP 112
F-91944 Les Ulis Cedex A
France

Phone: +33 (0) 1 69 18 75 75
Fax: +33 (0) 1 69 28 84 91

contact-edps@webofconferences.org

Additional copies of this publication are available from:

Curran Associates, Inc.
57 Morehouse Lane
Red Hook, NY 12571 USA
Phone: 845-758-0400
Fax: 845-758-2633
Email: curran@proceedings.com
Web: www.proceedings.com

TABLE OF CONTENTS

PREFACE

This conference contains the latest contributions presented in the 2018 International conference on Progress in Mechanical and Aerospace Engineering, which was held successfully in Hong Kong during August 10-12, 2018 with the help of our Local Chair: Prof Xingjian Jing from The Hong Kong Polytechnic University.

The PMAE2018 conference provided a platform to relevant scholars, researchers, and experts sharing and discussing a variety of Mechanical and Aerospace Engineering topics including Aerospace Mechatronics, Avionics Systems, Aerospace Communications, Aerospace Management and so on. This year, PMAE2018 attracted relevant representatives from different countries to present topics in their professional field.

The conference was divided in two parts: Morning Lectures with keynote speeches and Afternoon Sessions focused on author's presentations. All the participants valued this opportunity and were involved in research discussions. After the presentation sessions, the participants visited Prof Jing's Lab and other labs at the Hong Kong Polytechnic University which took more than 1 hour and the participants were impressed by the advanced instruments and equipment available for research and the research achievements of the various research groups.

We would like to thank all participants for their contributions to the PMAE2018 conference for their contributions to these Proceedings. Our special thanks go to our conference committees for their dedicated assistance in the overall organization of the conference. We also express our sincere thanks to Prof Xingjian Jing for his help in facilitating the conference rooms reservations. It was great to hold the conference in The Hong Kong Polytechnic University.

We are looking forward to the next PMAE and hope it will be even more interesting and it will attract a large number of contributors.

Prof. Dr. Ridha Ben Mrad, University of Toronto, Canada

2018-08-21

edp sciences

MATEC Web of Conferences

Statement of Peer review

In submitting conference proceedings to MATEC Web of Conferences, I certify to the Publisher that I adhere to the **Policy on Publishing Integrity** of the journal in order to safeguard good scientific practice in publishing.

1. All articles have been subjected to peer review administered by the proceedings editors.
2. Reviews have been conducted by expert referees, who have been requested to provide unbiased and constructive comments aimed, whenever possible, at improving the work.
3. Proceedings editors have taken all reasonable steps to ensure the quality of the materials they publish and their decision to accept or reject a paper for publication has been based only on the merits of the work and the relevance to the journal.

Title, date and place of the conference

2018 International conference on Progress in Mechanical and Aerospace Engineering (PMAE2018)
Hong Kong August 10-12, 2018

Proceedings editor(s):

Prof. Wei-Hsin Liao, The Chinese University of Hong Kong, Hong Kong

Prof. John Mo, Royal Melbourne Institute of Technology, Australia

Date and editor's signature

Analysis and Study of Heat Transfer Resistance Inside and Outside the Reentry Capsule

Chen Yang, Ning Xianwen, Su Sheng

Beijing Institute of Spacecraft System Engineering, Beijing, 100094

Abstract: According to the heat transfer characteristics inside and outside of the lunar-earth high-speed reentry capsule, a typical calculation model of heat conduction in external thermal protection system(TPS) coupled with internal radiation was established. The thermal properties of thermal resistance inside and outside the reentry capsule were analysed. The effects of thickness of the TPS, surface conditions and atmospheric pressures on the temperature were further explored. The results showed that atmosphere pressure was necessary to be controlled under 10Pa to ensure the safety temperature of the equipment and pipe. Based on the critical pressure, the configuration was optimized. The results provide detailed data for the system design of the lunar exploration, and also provide a reference for the thermal design of the atmospheric reentry spacecraft.

1 Introduction

In order to prevent the pneumatic heat eroding the inside wall in the return process, the outer surface of lunar-earth high-speed reentry capsule (reentry capsule) was fully coated with thermal protection system(TPS)[1]. So the enclosure formed a closed chamber, and the internal and external gas environment was unsmooth, and stated in a medium vacuum environment for long time.

The maximum heat insulation design[2] is used inside and outside the reentry capsule. Figure 1(a) shows the overall shape of the reentry capsule. The outer mounted TPS is different from the recoverable satellite[3] manned return module[4], the reentry capsule is thicker and the thickness range is 28~66.5mm.The thermal resistance of the thermal protection has a great influence on the thermal resistance of the whole re-entry capsule. The metal side wall and beam are covered by multilayer heat insulation (MLI) inside to ensure the small heat leakage in the in-orbit storage stage and to restrain the high enthapy during the return process. The external and internal thermal resistance affect each other. Therefore, it is important to analyze the influence of thermal resistance of each layer on the temperature level of the whole reentry capsule.

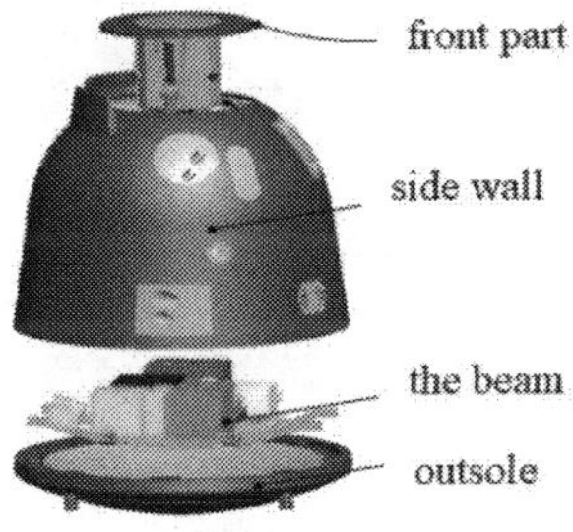

(a).overall shape of reentry capsule

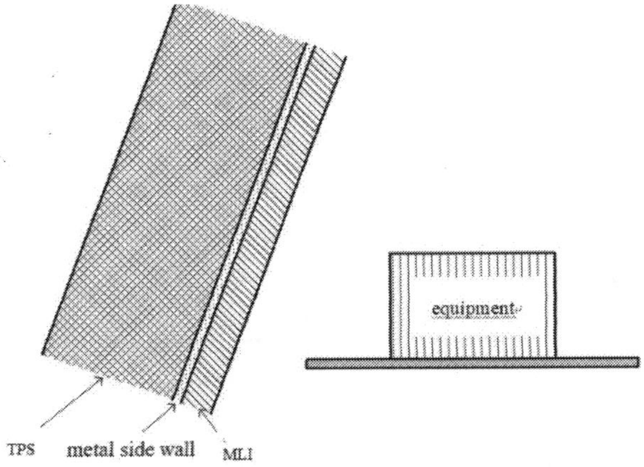

(b).physical model

Figure 1. Configuration of reentry capsule.

2 Reentry Capsule Heat Transfer Model

2.1 Physical Model

Heat transfer resistance inside and outside the reentry capsule includes thermal resistance of TPS, thermal resistance of metal structure, multilayer radiation resistance and beam in medium vacuum environment, thermal resistance of equipment, etc. Figure 1(b) shows the installation of the side wall and the equipment. The TPS is installed on the aluminum plate. Inside of the aluminum plate is covered with MLI, the surface of the equipment is blackened, and the MLI and equipment are radiative heat transfer. The pipe diameter Φ 6mm, pipe clamps are made of non-metallic polyimide material and pipe mounting seat is aluminum base, welding with side wall.

2.2 Mathematical Model

The physical heat transfer process can be simplified as follows:

(1) The convection heat transfer can be neglected, due to the small space in the device.
(2) The influence of cable and cable support on equipment occlusion is not considered.
(3) The thickness of the metal side wall is only 0.5 mm. It is considered that the temperatures of inner and outer layer of the MLI are equal.

Based on the above analysis, mathematical model is established as follows,

1) Temperature outside the thermal insulation structure

$$m_f C_{pf} \frac{dT_f}{d\tau} = Q_f + K_{af} A_h (\mathrm{T}_a - \mathrm{T}_f) - \varepsilon_f \sigma A_h T_f^4 \quad (1)$$

2) Multilayer temperature

$$m_a C_{pa} \frac{dT_a}{d\tau} = K_{ah} A_h (\mathrm{T}_h - \mathrm{T}_a) - K_{af} A_h (\mathrm{T}_a - \mathrm{T}_f) \quad (2)$$

3) Multi-layer inner temperature

$$m_h C_{ph} \frac{dT_h}{d\tau} = K_{ah} A_h (\mathrm{T}_h - \mathrm{T}_a) - \frac{\sigma A_{nr}}{1/\varepsilon_n + A_n (1/\varepsilon_h - 1) A_h} (T_n^4 - T_h^4)$$

(3)

4) Equipment temperature

$$m_n C_{pn} \frac{dT_n}{d\tau} = Q_n - Q_W - \frac{\sigma A_{nr}}{1/\varepsilon_n + A_{nr} (1/\varepsilon_h - 1) A_h} (T_n^4 - T_h^4) \quad (4)$$

m_n is quality of equipment, m_a is MLI bulkhead side mass, m_h is MLI inside side mass, m_f is quality of the TPS, Q_f is the absorbed external heat flux, C_{pn} is specific heat capacity of equipment, C_{ph} is MLI chamber film heat capacity, C_{pa} is MLI bulkhead side heat capacity, C_{pf} is heat capacity of TPS, A_{nw} is contact area between equipment and mounting plate, A_h is MLI area, A_f is the TPS area,ε_n is surface emissivity of equipment, ε_h is MLI emissivity, ε_f is TPS emissivity, K_{ah} is MLI heat transfer coefficient, K_{af} is internal and external heat transfer coefficient of the TPS.

3 Analysis of the Thermal Characteristics of the Reentry Capsule

3.1 Analytical Model Parameters and Boundary Conditions

According to the characteristics of the overall configuration of the reentry capsule, the structural design and the system parameters, the analytical model parameters are determined, as shown in table 1.In the analysis, the cruising section with the worst low temperature of the reentry capsule is selected for analysis. The material parameters are taken at the beginning of the life, and the external heat flux is applied to the ship. The laser IMU and the typical pipe is selected for analysis and research.

Table 1. Analysis of model parameter values.

Code	Values	Code	Values	Code	Values
m_n	7.65kg（laser)	A_{nr}	0.185m^2	Q_f	128W
$m_{a(h)}$	50.4g	A_{nw}	0.042m^2	Q_n	16W
m_f	25.2Kg	A_h	5.04m^2	C_{pn}	835J/Kg.K
m_w	0.93kg	A_f	5.04m^2	C_{ph}	600J/Kg.K
K_{af}	0.00524W/m^2.K	A_w	0.15m^2	C_{pa}	600J/Kg.K
ε_w	0.1	ε_n	0.88	C_{pf}	1.4×10^{-3}J/kg.K
ε_f	0.87	ε_h	0.1	C_{pw}	900J/Kg.K

3.2 Results and Analysis of Influencing Factors

All painted white paint on the outside of the reentry capsule, the required heat of the propulsion system pipe is 16W2.In cruise attitude flight, only the top surface is radiated by external heat flux, and the absorption of external heat flux is relatively small.During cruise flight, the two IMU should meet the requirements of starting temperature 0℃, and the pipe temperature(Tpipe) should be maintained above 5℃.The steady-state results are as follows: the temperature of IMU (TIMU) is 2.9 ℃, and the temperature of typical pipe is 5.9 ℃.

In the thermal resistance inside and outside the device, the MLI thermal resistance and the thermal resistance of the TPS are the two most important thermal resistances.The decrease of true test of MLI environment has a great influence on MLI performance56. The thermal resistance of TPS mainly depends on the thickness of the TPS. Therefore, the influence of MLI vacuum degree and thermal insulation thickness on the temperature of electronic equipment and pipe is evaluated by setting up the analysis condition of the influence of the vacuum degree and the thickness of the TPS, respectively.

Influence of thickness of the TPS

Under other conditions, the thickness of the TPS is reduced to 3~16mm from the original 33mm, and the results are shown in table 2. It can be seen that when the thickness of the TPS is reduced to 11mm, the temperature of the laser IMU will be less than 0℃.

Table 2. Thickness of TPS.

Working condition	Thickness/mm	T_{IMU} /℃	T_{pipe} /℃
Case 1	0.033	2.9	5.9
Case 2	0.016	0.7	5.8
Case 3	0.011	-0.1	5.6
Case 4	0.0055	-0.4	5.2
Case 5	0.0003	-1.2	4.6

(1) Vacuum effect

T_{IMU} and T_{pipe} obtained by steady-state calculation under different vacuum conditions are shown in table 3.With the pressure increasing, T_{IMU} and T_{pipe} is reduced. When the pressure is greater than 10Pa, T_{pipe} is lower than 5℃. When the pressure is more than 50Pa, T_{IMU} is lower than 0℃.The temperature sensitivity of propulsion pipeline to vacuum degree is stronger than that of equipment because of the influence of multiple layers of bulk coating and multi-layer of side wall.

Table 3. Different vacuum analysis conditions.

Working condition	Internal pressure/Pa	T_{IMU} /℃	T_{pipe} /℃
Case 6	0.001	15.3	9.4
Case 7	0.01	9.1	7
Case 8	1	1	6.4
Case 9	10	2.9	5.9
Case 10	50	0.6	4.4
Case 11	100	-2.6	3.2

(2) Effect of MLI surface condition

The influence of infrared emissivity is considered in the MLI of the device, and the polyester film, polyimide film and F46 face film are set up respectively.

It can be seen from table 4 that the heat transfer between the equipment and the bulkhead increases and the equipment temperature decreases with the increase of the MLI emissivity in the device.The reentry device is in the cruise section and the reentry section is designed for heat insulation. During the two flights, the heat control power is less. Therefore, the selection of polyester film on the multi-layer surface is the most favorable.

Table 4. MLI surface condition.

Working condition	MLI surface shape	T_{IMU} /℃
Case 12	Polyester film, emissivity 0.12	2.9
Case 13	F46 surface, emissivity0.68	-13.9
Case 14	Polyimide film, emissivity0.88	-15.8

4 Structural Optimization

According to the results in Table 3 , if T_{pipe} can be controlled above 5℃, the internal pressure value of the device cannot be greater than 10Pa.Since the pressure inside and outside of the return device is not smooth, it is necessary to design a suitable pressure relief hole without affecting the pneumatic, ablation and measurement and control, and to combine the idea of unified design of heat prevention and heat control grooming[78].Therefore, two Φ 6mm holes are designed at the front.The designed structure verifies the pressure relief ability of the through hole in the vacuum tank. After 7 hours of vacuum pumping, the pressure in the device decreases to 7.1 Pa.
In the heat balance test of the reentry capsule, two vacuum meters were installed at the inner side wall of the injector, and the vacuum degree in the device was monitored. The pressure reduced to 8.9Pa in 6 hours. During the thermal test, the vacuum degree inside the device was higher than 0.12Pa.

5 Conclusion

Based on equilibrium thermal mathematical model, the matching between thermal resistances was analyzed. The pressure relief of different vacuum degree was investigated by numerical simulation. The influence of different thickness of the TPS and different surface condition on the temperature level in the device can be concluded as follows.(1)The effect of vacuum degree on the MLI of the reentry capsule was more serious. If the vacuum degree was greater than 10Pa, T_{pipe} would not meet the requirements of the index. (2)The thermal resistance of the TPS was large, when the average thickness of TPS was reduced from 33mm to 11mm, T_{IMU} would be lower than 0 ℃, which can not meet the starting temperature index.(3)The MLI surface condition has a great influence on the temperature of the device. If the F46 or polyimide film was chosen, the temperature of the device would drop sharply, which can not meet the temperature requirement. Therefore, polyester film is beneficial to the insulation of cruise section and reentry section.

6 Reference

1. Yang M F, Zhang G, Zhang W, et al. Sci Sin Tech, 2015, 45:111–123.
2. Ning X W, Su S, Chen Y, et al. Sci Sin Tech, 2015, 45:145–150.
3. Dong Y Z. China Association of Science and Technology Annual Conference, Beijing, 2006.
4. Huang J R. Manned Spaceflight. 2012.2.
5. Chen Y, Ning X W, Su S, et al. Sci Sin Tech, 2015, 45: 263–267.
6. Liu W, Wen Y P. National Fifth Conference on Space Thermal Physics Huangshan, 2000
7. Wu G T. Spacecraft Engineering, 2009, 18(4):13-19.
8. Li W J, Liu F, Dong Y Z, Li X L. Spacecraft Engineering, 2013, 22(3):113-118.

Migration and Evaporation Characteristics of Kerosene Droplet in Supercritical Environment

Hailong Wu [1], Wansheng Nie [1], Zhi Zheng [1] and Yu Liu [1]

[1] *Space Engineering University, 101416Beijing, China*

Abstract. A one-dimensional full transient droplet evaporation model was established under consideration of factors such as high pressure vapour-liquid equilibrium, high-pressure physical property corrections, gas phase dissolution, and shift of interface. The finite volume method was used for discretization to study the migration and evaporation characteristics of the surrogate fuel for kerosene, which was constituting of (mass fraction)80% n-decane and 20%1,2,4-trimethylbenzene, under supercritical conditions. The results show that, under supercritical conditions, the higher the temperature and the pressure, the easier and the sooner the supercritical migration occurs. Before the supercritical migration occurring, there was obvious boundary between the gas and liquid phases. The mass fraction of component was discontinuous, and the gradient of temperature near the interface was large. After the supercritical migration occurring, the surface of the droplet disappeared, there was no obvious boundary between the gas and liquid phases, and the distribution of the components mass fraction and temperature were continuously. With the increase of the initial temperature of the droplet, the time of the supercritical migration was greatly advanced, the rise rate of the droplet surface temperature increased, and the phenomenon of endothermic expansion no longer appeared.

1 Introduction

With the rapid development of the aerospace industry around the world, people need to send more and more things into space for various scientific experiments, planetary exploration, and the establishment of space stations. Therefore, rocket engines with large thrust and high efficiency are urgently needed.

At present, the heavy-lift rocket engine is mainly on liquid oxygen/kerosene rocket engines. In order to improve the performance of the liquid oxygen/kerosene rocket engine, people continuously increase the pressure of the engine combustion chamber to improve the combustion efficiency. However, with the increase of the pressure, the working environment of the combustion chamber is completely supercritical. The related theories of injection, atomization, evaporation, mixing, and combustion in the past under low pressure and subcritical conditions have not been applied, especially, the process of evaporation and combustion [1]. Therefore, it is necessary to establish the theory of droplet evaporation and combustion under supercritical conditions.

The experimental research of kerosene droplet evaporation under supercritical environment is mainly on hanging-droplet method. GHASSEMI H[2] studied the evaporation process of kerosene droplet in the pressure range of 0.1 to 3 MPa and the temperature range of 500 to 1000 °C using the hanging-droplet method. It was found that with the increase of pressure, the evaporation rate increases first and then decreases at high temperture. At low temperatures, when ambient pressure increased, evaporation rates monotonically increase. Jin Le et al. [3] studied the effect of the droplet initial temperature on the supercritical evaporation of RP-3 aviation kerosene by experiments, founding that the higher the initial temperature, the larger the droplet diameter when the droplet appeared supercritical scattering spots. In terms of numerical simulation, the single-component surrogate fuel for kerosene is mainly used for studying. Li Yunqing et al.[4] studied the process and conditions of the surface migration in the supercritical evaporation process, based on the actual gas state equation and high-pressure vapour-liquid equilibrium. Li Pengfei et al.[5] studied the variation of gas solubility, evaporation rate and droplet life under subcritical and supercritical conditions, which was based on OpenFOAM and using real fluid model and state equation method to calculate the multi-component vapour-liquid equilibrium under high-pressure. B. Balaji [6] studied how the temperature and pressure affect droplet survival time of n-dodecane droplet in the subcritical and supercritical environment. There are also some multi-component supercritical evaporation models, such as Zheng Li[7] established a multi-component evaporation model under high-temperature and high-pressure, and studied of the evaporation process of dimethyl ether and liquefied petroleum gas mixture, but the supercritical model is very simple.

A one-dimensional full transient droplet evaporation model has been established and a simulation program, which can be both used for single-component and bicomponent droplet evaporation, was written using Fortran language. And the effectiveness of the program was verified. In the meantime, a bicomponent surrogate fuel for kerosene was found, and the transcritical characteristics of the kerosene droplet and the effect of initial droplet temperature on the transcritical characteristics of the kerosene droplet under the supercritical environment was studied by simulation.

2 Modeling

2.1 Basic assumptions

In an actual liquid rocket engine, the shape of droplet formed by atomization is not completely spherical, and the environment around the droplet are relatively complicated. So the droplet model under actual conditions is simplified for the convenience of the simulation:

(1) Ignore the airflow velocity and pressure fluctuations around the droplet.

(2) Ignore the effects of the gravity and natural convection.

(3) Ignore the shape change of droplet.

(4) Ignore Dufour and Soret Effect of Gas Phase.

(5) Ignore the effects of thermal radiation.

A one-dimensional full transient droplet evaporation model was established based on the above basic assumptions.

2.2 Governing equations

The governing equations of the liquid and gas phase are basically in the same form, and after the entering of the supercritical evaporation stage, the liquid phase and the gas phase no longer separate. Therefore, the unified governing equations were established for the liquid phase and the gas phase.

$$\frac{\partial \rho}{\partial t}+\frac{1}{r^2}\frac{\partial}{\partial r}(r^2\rho v)=0 \tag{1}$$

$$\frac{\partial(\rho Y_i)}{\partial t}+\frac{1}{r^2}\frac{\partial(r^2\rho v Y_i)}{\partial r}=\frac{1}{r^2}\frac{\partial}{\partial r}(r^2\rho D\frac{\partial Y_i}{\partial r}) \tag{2}$$

$$\frac{\partial(\rho C_p T)}{\partial t}+\frac{1}{r^2}\frac{\partial(r^2\rho v C_p T)}{\partial r}=\frac{1}{r^2}\frac{\partial}{\partial r}(r^2\lambda\frac{\partial T}{\partial r}) \tag{3}$$

In the formula, r represents the distance from a point to the center of the droplet in the physical plane(m); t represents the time(s); Y_i represents the mass fraction of the components; ρ represents the density(kg•m^{-3})(gas phase: ρ_g, liquid phase: ρ_l); T represents the temperature(K)(gas phase: T_g, liquid phase: T_l); v represents the velocity(m•s^{-1})(gas phase: v_g, liquid phase: v_l); D represents the self-diffusion coefficient(m^2•s^{-1}) (gas phase: D_g, liquid phase: D_l); C_p represents the specific heat at constant pressure(J•kg^{-1}•K^{-1}) (gas phase: C_{pg}, liquid phase: C_{pl}); λ represents the coefficient of heat conduction(W•m^{-1}•K^{-1}) (gas phase: λ_g, liquid phase: λ_l).

2.3 Initial and boundary conditions

2.3.1 Initial conditions

At the time of t=0,

$$T_l=T_{l0}\,,\ T_g=T_\infty\,,\ P_g=P_l=P_\infty$$

T_{l0} represents the initial temperature of the droplet, T_∞ and P_∞ represent the temperature and pressure at the boundary of infinity. At the initial conditions, there are only fuel components inside the droplet, and only ambient gas components in the gas phase.

2.3.2 Subcritical boundary conditions

In the infinity of gas phase:

T=T_∞、Y_i=0 or 1(it is 0 when the i represents the fuel component and it is 1 when the i represents the nitrogen).

At the center of the droplet:

$$\frac{\partial T}{\partial r}=0\,,\ \frac{\partial Y_i}{\partial r}=0\,,\ \frac{\partial v}{\partial r}=0\ .$$

At the surface of the droplet:

$$T_{g,s}=T_{l,s}=TS\,,\ \phi_i^l X_i^l=\phi_i^g X_i^g\ .$$

Of which, $T_{g,s}$ and $T_{l,s}$ are the gas-phase and liquid-phase temperatures of the droplet surface, respectively. TS is temperature of the droplet surface. X_i^g and X_i^g are respectively the gas-phase and liquid-phase fugacity coefficient of the droplet surface, which were Obtained by the Peng-Robinson equation of state [8].

2.3.3 Supercritical boundary conditions

When the temperature and pressure at the gas-phase of droplet surface exceed the critical value of the mixture, it is considered that the supercritical state has reached. At this time, the surface tension of the droplet is reduced to zero, the surface of the droplet disappears, and the thermal properties between the gas and the liquid phase tend to continuous [9]. Therefore, when the surface of the droplet reaches a supercritical state, the gas-liquid phase is no longer distinguished, and only the boundary conditions at the centre and the surface of the droplet are defined, and its definition is the same as the subcritical state.

2.4 Numerical computation method

The surface of the droplet changes with the evaporation process, if the control equation is directly discretized and solved, it is necessary to repartition the calculation grid at each time step. In order to avoid it, using the formulas (7) and (8) to convert the coordinate [8] from the physical plane to the calculation plane to make the droplet surface always fixed at the position where the calculation coordinate is 1.

Spatial position transformation of two coordinates:

$$\zeta = r / R_S \tag{7}$$

Velocity transformation of two coordinates:

$$w = v - \zeta \dot{R}_S \tag{8}$$

Of which, ζ represents the distance from a point to the center of the droplet in the calculation plane, R_s represents the radius of the droplet(m), $\dot{R}_S$ represents the change rate of the droplet radius.

Since the flow field parameters near the surface of the droplet varies greatly, the grid is divided into non-uniform grids near the surface of the droplet to make the grid refinement, thereby improving the calculation accuracy of the flow field near the droplet surface.

The finite volume method is used for the discrete method of the governing equations with coordinate conversion, and the control volume boundary is taken as the centre of the adjacent grid node [10].

3 Model verification

In order to test the effectiveness of the calculation model and the program, the evaporation process of n-heptane droplet with a diameter of 50 μm under different pressures and temperatures in a nitrogen atmosphere was calculated, and the results were compared with the results of the literature [9]. As shown in Figure 1.

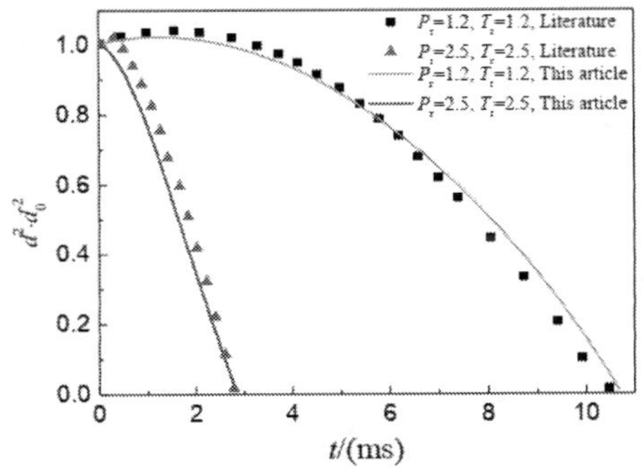

Figure 1. Simulation result comparison and verification

In the figure 1, *Pr* represents the ratio of ambient pressure to n-heptane critical pressure. T_r represents the ratio of ambient temperature to n-heptane critical temperature. d and d_0 represents the diameter and initial diameter of the droplet, respectively. It can be seen that the calculation results in this paper is in good agreement with the calculations in the literature under the conditions not only of lower temperature and pressure but also of higher temperature and pressure. The correctness of the model has been verified.

4 Calculation results and analysis

4.1 The selection of surrogate fuel for kerosene

At present, the model number of aviation kerosene mainly used in our country is RP-3. The composition of RP-3 aviation kerosene was analysed by gas chromatography-mass spectrometry in literature [11]. The results show that RP-3 is composed of 42.330% paraffins, 21.348% cycloparaffins, 24.045% benzenes, 8.184% naphthalenes, and 4.093% others. RP-3 aviation kerosene is a complex mixture. Now, it is impossible to directly perform the simulation calculation of evaporation process of the complex mixture droplet under the supercritical environment. This article uses the bicomponent surrogate fuel for RP-3 aviation kerosene proposed in [12] and [13]: (mass fraction) 80% n-decane and 20% 1,2,4-trimethylbenzene. Its main thermodynamic properties are shown in Table 1. There are several main reasons for using this solution:

(1) The simulation of the bicomponent surrogate fuel is easier to realize.

(2) Compared to single-component surrogate fuel for kerosene, bicomponent surrogate fuel for kerosene can simulate the mixture characteristics of kerosene.

(3) From analyzing the components of kerosene, it can be seen that alkane and benzene series are the main components. Selecting n-decane and 1,2,4-trimethylbenzene can be a good substitute for the physical and chemical properties of alkanes and benzenes, and laying a good foundation for the subsequent simulation of kerosene droplet combustion.

Table 1. The components and thermodynamic properties of the bicomponent surrogate fuel for kerosene.

Component	mass fraction (%)	Critical pressure (atm)	Critical temperature (K)	Boiling point (K)
n-decane	80	20.8	617.7	447.3
1,2,4-trimethylbenzene	20	31.9	649.1	442.5

4.2 The evaporation of droplet in a supercritical environment

In this paper, a bicomponent surrogate fuel, which was composed of n-decane and 1,2,4-trimethylbenzene, for kerosene droplet is used in the calculation. The initial diameter of the droplet d_0 is 100μm, and the initial temperature of the droplet T_{l0} is 300K. The surrounding gas is nitrogen.

In the figure 2, it shows the variation of the relative diameter square of the droplet and the surface temperature of the droplet under three working conditions P=80atm T=1200K, P=100atm T=1400K and P=120atm T=1600K. In the initial stage of droplet evaporation, the surface temperature of the droplet rapidly rises, and the droplet diameter increases. This is mainly because of the large temperature differential between the droplet initial temperature and the ambient temperature. The droplet rapidly absorbs heat and expands, but at this time almost all of the absorbed heat is used to increase the temperature of the droplet, causing the evaporation rate of the droplet to be close to zero, the droplet to be thermally expanded, and the diameter of the droplet to increase. As the temperature differential between the droplet surface and the ambient temperature decreases, the droplet enters the rapid evaporation stage. Although the surface temperature of the droplet increases, the increase rate is little, and the droplet diameter rapidly decreases until the droplet evaporation finishes or the super critical migration occurs.

With the increase of the ambient pressure and temperature, the temperature rising rate of the droplet surface in the initial stage of evaporation increases, but the maximum temperature that can be achieved is almost the same. The decreasing rate of the square of the droplet diameter is also increasing. At the initial stage of evaporation, the phenomenon of heat absorption and expansion of droplet becomes less and less obvious. At the same time, the higher the pressure and temperature are, the earlier the supercritical migrate of the droplet occurs.

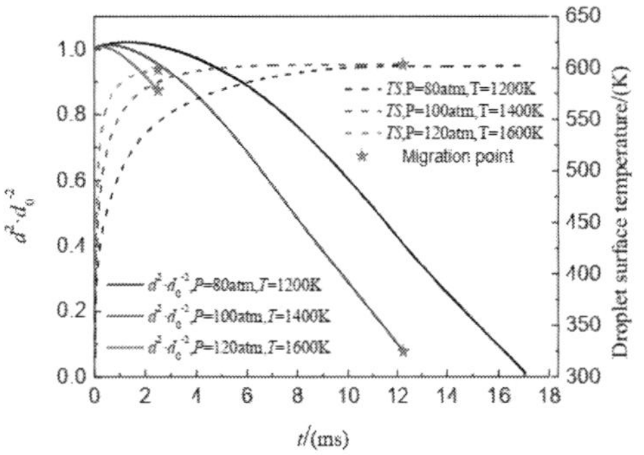

Figure 2. Variation of droplet relative diameter square and droplet surface temperature

4.3 The evaporation characteristics of droplet before and after supercritical migration

In the figure 3, it shows the distribution of the mass fraction and temperature along the radius of each component of the droplet before and after the supercritical migration at P = 120 atm and T = 1600 K. Under the condition, the droplet life is 13.61 ms, and the migration time is 2.51 ms. The time of 1.25 ms and 7.51 ms are selected, which is the middle moment of two stages before and after the supercritical migration.

It can be seen that the mass fraction of n-decane and 1,2,4-trimethylbenzene inside the droplet changes very little before the occurrence of supercritical migration, which is basically the initial value, and the mass fraction of nitrogen is almost zero. In the vicinity of the droplet surface, there is an obvious discontinuity, that is to say the mass fraction of the components is not continuous at the interface. At the same time, the liquid phase temperature is also low, high temperature only occurs near the droplet surface. Although the temperature at the gas-liquid interface is continuous, the temperature gradient is large.

After the occurring of the supercritical migration, the mass fraction of n-decane and 1,2,4-trimethylbenzene in the original liquid phase of the droplet decreases significantly, and the mass fraction of nitrogen increases. Along the radius r, the mass fraction of the components is continuously distributed and there is no interruption. At the same time, the temperature inside the original phase of the droplet increased and the temperature gradient decreased significantly.

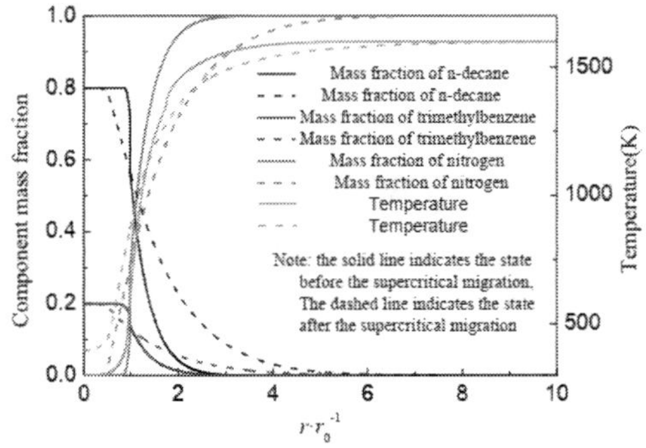

Figure 3. The distribution of component mass fraction and temperature

4.4 The effect of initial droplet temperature on the supercritical migration

Figure 4 and 5 are the time-dependent curves of the droplet relative diameter square and the droplet surface temperature under the initial conditions of P=100atm T=1400K, respectively. It can be seen that under the same environmental conditions, the initial temperature of the droplet has a great influence on the evaporation characteristics of the droplet, especially the supercritical migration characteristics. As the initial temperature of the droplet increases, the point of supercritical migration of the droplet occurs more and more earlier, and which even occurs immediately at the beginning of evaporation.

In figure 4, as the initial temperature of the droplet increases, the endothermic expansion phase of the droplet evaporation process becomes more and more unapparent. When the droplet initial temperature reaches 450K, there is no endothermic expansion phase. At the same time, it can be seen from figure 5 that with the increase of the initial droplet temperature, the rise rate of droplet surface temperature significantly increases although the initial temperature differential between the gas and liquid phases decreases. The above phenomenon occurs mainly because the initial temperature of the droplet becomes closer to the boiling point of the mixture, as the initial temperature of the droplet increases, the more easily the gasification occurs.

Therefore, it can be seen that in the actual engine working process, the combustion efficiency can be improved by increasing the initial temperature of the injected fuel to accelerate the evaporation rate of the atomized droplet. Especially in the ignition phase, accelerating the evaporation rate can greatly reduce the ignition delay.

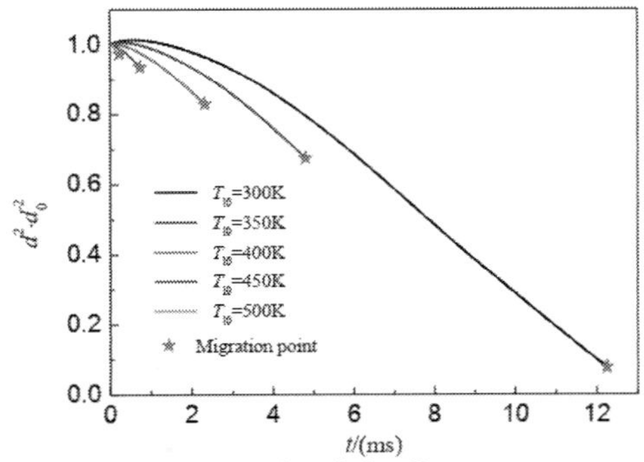

Figure 4. Variation of droplet relative diameter square

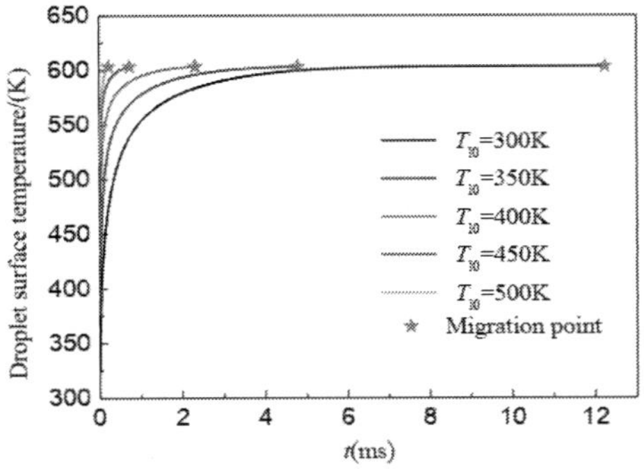

Figure 5. Variation of droplet surface temperature

5 Conclusion

The paper first establishes a simulation calculation program for the droplet non-steady-state evaporation model in supercritical environment and verifies the validity of the calculation program, then searched for a bicomponent surrogate fuel for RP-3 aviation kerosene. Finally, the evaporation characteristics, supercritical transport characteristics and effects of initial droplet temperature on evaporation and migration characteristics in supercritical environment were studied by simulation calculations. The results not only lay the foundation for the simulation research of kerosene droplet combustion in the supercritical environment, but also provide some reference for explaining the mechanism of ignition and unstable combustion in liquid oxygen/kerosene rocket engines. The main conclusions are as follows:

(1) Under supercritical conditions, as the ambient temperature and pressure increase, the droplet are more prone to occurring supercritical migration and the earlier the supercritical migration occurs.

(2) Before the occurrence of supercritical migration, there were obvious boundaries between the gas and liquid phases, the distribution of mass fraction of each component was discontinuous and the gradient of temperature near the interface was large. After the supercritical migration occurs, the droplet surface disappears, there is no obvious boundary between the gas and liquid phases, and the mass fraction and temperature of each component are continuously distributed.

(3) With the increase of the initial temperature of the droplet, the supercritical fluid migration point advances significantly. The rise rate of the droplet surface temperature increases, and the droplet no longer displays the phenomenon of thermal expansion.

References

1. F. C. Zhuang, National University of Defence Technology Press. M(1995)
2. H. GHASSEMI, S. W. BAEK, Q. S. KHAN, Combustion Science and Technology. J. 178. 1669-1684(2006)
3. L. Jin, W. Fan, Z. Zhou, et al. Journal of Combustion Science and Technology. J. 22. 269-275(2016)
4. Y. Q. Li , H. N. Wang, W. Chen, Journal of Combustion Science and Technology. J. 16. 287-294 (2010)
5. P. F.Li, F. P. Lei, K. Wang, et al. Acta Aeronautica et Astronautica Sinica. J. 39. 121764(2018)
6. B. Balaji, Vasudevan Raghavan, K. Ramamurthi, et al. Physics of Fluids. J. 23.063601(2011)
7. L. Zheng, Wuhan University of Technology. D(2014)
8. R. Ruszalo, The University of Ottawa. D(1990)
9. P. He, School of Transportation Science and Engineering Beihang University. D(2010)
10. F. J. Wang, Tsinghua Universit Press. M(2004)
11. Z. Zhou, W. Fan, L. Jin, et al. Journal of Propulsion Technology. J. 37. 1422-1430(2016)
12. H. R. Wang, Gas Turbine Experiment and Research. J. 28. 19-23(2015)
13. S. Honnet, K. Seshadri, U. Niemann, et al. Proceedings of the Combustion Institute. J. 32. 485-492(2009)

Design and Manufacturing Process of Micro Air Vehicles

Mateusz Rakiec , Aleksandra Kwiecien , Rafal Muchowski and Bartosz Kazmierski

Rzeszow University of Technology, Powstancow Warszawy Av. 12, 35-959 Rzeszow, Poland

Abstract. This paper shows shortly the designing, optimization and manufacturing process of light and compact, carbon fiber MAV. It also covers testing and results. The main purpose was to make an innovative prototype of UAV which is easy to transport, assembly and carry as much high volume payload as possible. The optimization process started from defining the constrains. To gain information needed to choose the best solution and perform optimization, authors made many tests. The material, which was chosen, had to be as light and durable as possible – this characteristics were later checked in tensile test of carbon fiber samples. The main optimization algorithm do initial research about the rest parameters of aircraft. The point was to narrow the range of analyzed chord's values and find the best motors. For better performance the original airfoil was developed. It works perfectly in low Reynolds number area and has perfect characteristics for high-lift airfoils, compared to similar. It was later checked in CFD. The special manufacturing process of carbon fiber airplane was designed. After all the construction was tested. The successful flight tests confirmed that design process was appropriate and done well. All theoretical assumptions were also right.

1 Initial Considerations

The main objective of this paper is to design a small, light-weight unmanned aerial vehicle, capable of lifting the greatest possible payload compared to empty mass. It should carry high volume payload and be assembled quickly from the small package. The designing process of heavy UAVs is well developed these days. There are also plenty of publications about design mini unmanned air vehicles but they mainly cover the quadcopter and similar systems and those about fixed-wing constructions, which are very light, could be back-packed, assembled and deployed by just one person, are not well developed [1-2]. The most important aspects to find best solution which meet the criteria set above are:

- make an MAV which will be as light as possible but also durable,
- aircraft must be hand launched by one person and land on the ground
- construct the airplane, which will be packed in the box with small dimension and be easy to assemble
- design the aircraft as safe and simple as possible.

To make it possible, was planned:

- make a carbon fiber, single engine UAV with the lowest empty mass, which fits in the box,
- develop software to optimize the construction for best performance,
- find appropriate materials – light but durable and design best manufacturing process.

The main problem to discuss was the dimension constraints imposed by box dimensions. The optimization software was developed to deal with the problem and to analyze various range of size configurations very thoroughly.

2 Final Conception

The main limitations of constructions are dimensions of packing box, assembly time and payload's requirements. Also that the maximum dimension of single piece caused by box is 35cm, maximum payload size is 10 cm x 10cm x 35cm, wing chord between 9,9 cm and 25,8 cm and maximum angle of attack 8°.

The number of wing sections that could be possibly fitted inside the box is three. This impose the constraint of wingspan about 1m depending on dihedral. The wingspan limitation suggested using rectangular, no swept wing to obtain the largest surface, that seemed to be the key in high lifting capacity.

For best configuration selection the discriminators were defined with specified importance and differentiated from each other. Many ideas were considered and in the end conventional configuration was chose. It is light and easily controllable as long as designed properly, provides fast loading and assembly possibilities. Aircraft is also predictable and, due to extra stability, the payload might be loaded over the wing.

To find the dihedral configuration the CFD analysis was performed to assess whether the expected growth in induced drag, due to the shape, has enough small impact to remain rectangular wing assumption. The collected data by ANSYS software confirmed the benefit of above ascertainment. Two options of wing dihedral configuration were took into consideration. The analysis suggested application of double dihedral configuration for best lift to drag ratio and simple assembly.

When discussing the issue of payload layout and its fixing, many variants were took into consideration.

The choice was between the payload mounted over the wing and under it, with lowest wing area covered. Basing on the fact that landing vulnerability is more essential than maneuverability and avoiding the payload damage the payload mounted over the wing was chosen.

3 Airfoil Selection

The worst part with finding best airfoil was to deal with low Reynolds numbers area. Every known high lift coefficient characteristic wasn't appropriate for particular values of Reynolds number around 100 000, so the new one was designed [3-6]. After analysis in XFLR5 the RML (red line on Figure 1) was developed, which fits perfectly for the main purposes. It has linear $C_L(\alpha)$ ratio and, compared to other airfoils, higher critical angle of attack. This makes construction safer – smooth stall characteristics is placed at about 12°. Also airfoil's $C_L(C_D)$ ratio is much better and makes an aircraft more stable. Moreover the drag coefficient is lower. The figures show that in selected Re numbers area many airfoils' C_L vs C_D dependence is irregular, which is an adverse. It means that pilot could have problem with control an aircraft during take-off and landing. Whole research has proved that design process was right. It was later checked in ANSYS software.

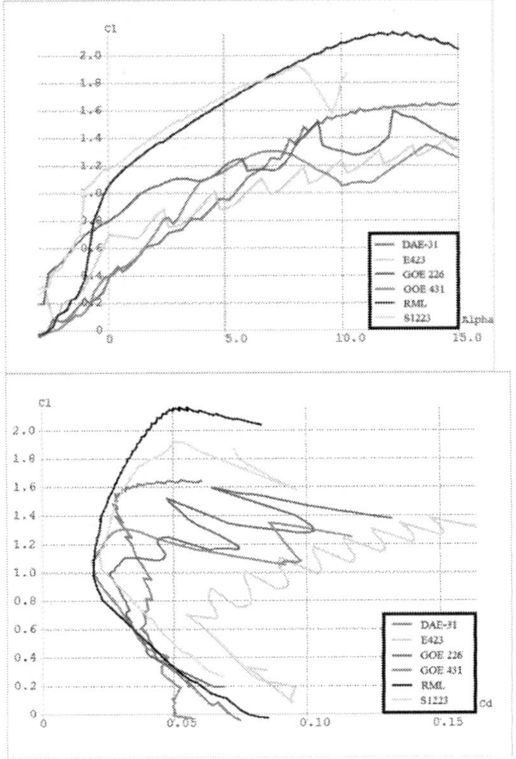

Figure 1. Airfoil comparison

4 Optimization

4.1 Initial Performance Considerations

The generic algorithm in Matlab software was developed to do initial research about what size of electric motor will probably be used and what the estimated range of chord dimension was. The additional constraint, resulting from the algorithm structure, was the limit for maximum motor. The input of the application was the airfoil deducted by aerodynamic research. The gradually increased variables were: wing's chord, an angle of attack and dynamic thrust approximation. The point was to narrow the range of analyzed chord's values and to decide what motors should be used in some detailed tests.

4.2 Optimization Process

Since the crucial data was collected, an exact size of wing and its angle of attack was calculated. The block diagram of second, final optimization software is in Figure 2.

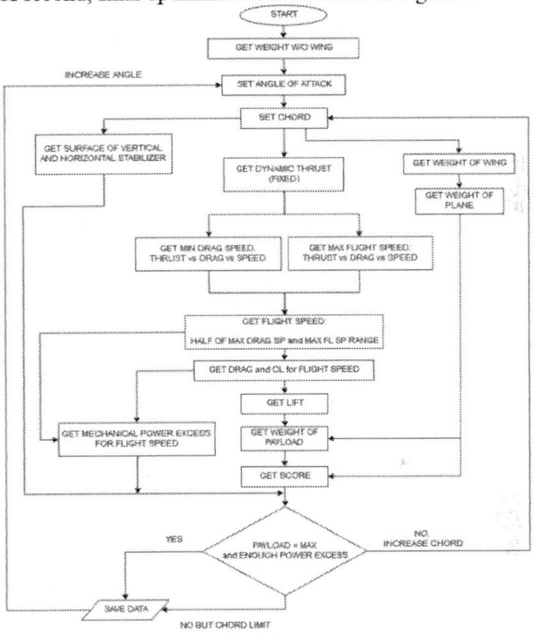

Figure 2. Final optimization loop

The variable list was divided into major variable - chord of wing, and the minor variable - angle of attack. Both of them were increased step by step to make sure to not omit any solution [7]. The algorithm helped to get information about the size of horizontal and vertical stabilizers. It was also necessary to arrange the way of flight speed deduction. The application calculated the corresponding values of C_l, C_d, Reynolds number and total drag depending on velocity, including parasite drag, induced drag, drag generated by fuselage, empennage, fittings and so on. The issue, which had to be solved, was what value of airplane's speed to be used for further calculations. The maximum flight speed and minimum drag speed were found by transforming drag and thrust data into mechanical power data [2]. The decision was to use the exact middle value from range between the minimum drag speed and maximum flight speed. Also next criteria was defined to have at least 20% of mechanical power excess, defined as the ratio of difference between the power available and power required to power required. The reason was to ensure possibility to climb or remain constant altitude of flight in poor weather conditions and to compensate possible manufacturing inaccuracies.

Determined value of flight speed, lift coefficient and wing surface were used to calculate the lift [8]. The difference between the lift and empty weight of the aircraft multiplied by safety factor, which was assumed to be equal to 0.9, expressed the weight of payload possible to carry by given parameters configuration. The application of safety factor results from the risk of momentary unpredicted flight issues and risk of calculation method deviation. The algorithm was checking whether the airplane could carry the predicted payload of weight 1.4 kilograms and whether there is at least 20% of mechanical power excess. Meeting these both conditions or lack of further values of chord from range 9.9-25.8cm made the data was saved on hard disk for analysis and angle of attack was increased. In such case the process was performed again. Otherwise, the chord was increased without rising an angle. The maximum optimization performance was reached for maximum angle of attack (8°) and chord 11cm but there was also growth for 5°. After taking into consideration that optimal angle of attack, according to the aerodynamic research, is 2.5° and basing on the above plots increasing this angle ensures higher score due to lower weight of empty MAV. Application of such great angle as 8° will make the flight hazardous because of potential risk of stall or shortage of controllability. The compromise between those two aspects, it was decided to choose 3.5° angle of attack and 16cm of chord. Longer chord will cause problems with placement of the airplane inside of the box. Low extension of the sought values should not bring a significant lose in competition results but will make predictability of flight higher. This extension was another, the third optimization precaution, since awareness of the risk of calculations deviation came from slight simplifications. An angle of attack of the wing was adjusted by the pilot during flight and lift force generated by horizontal stabilizer is changed by the deflection of elevator, that is why the suggested value of angle does not have to be strictly followed. However, deviation from calculated value cannot be too excessive because of additional growth in drag generated by fuselage and tail. That is why angle of attack is a secondary variable and is meant to be a clue for further design and analysis how much the change in the angle influences the maximum payload. Another purpose of this was to show particularly beneficial range of angles of attack if such exists.

5 Structural Design

By using composites it was possible to make light and durable spar. Spar flange is made of carbon fibre filaments and Rohacell inside. Because carbon roving has enormous strength, classic calculations were useless. Minimum cross section of spar was defined by the means of bending moment (taken from XFLR5) that was converted to tensile and compressive forces in the box spar in every segment [9]. The stiffness of a material is the most important in this technology so the tensile testing on Zwick/Roell Z100 machine were performed. Many different carbon fibre samples were made and tested to find those which meets load requirements for spar flange.

6 Analysis

6.1 Dynamic & Static Stability

The static stability was tested with full payload onboard in AVL, at airspeed predicted for landing phase equals to $v = v_{stall} * 1.2 = 8.6\frac{m}{s}$. It showed that static margin of 17% was enough for good longitudinal static stability. Dynamic stability was analyzed in a similar manner, with fully loaded aircraft's moments of inertia obtained via CAD model. Every point of the system was in upper left quadrant of a plot, with negative real components, which means that each of the five modes are stable.

6.2 Horizontal & Vertical Stabilizer

The next point was to define the type of fuselage. Since there is limited space it was decided to choose straight-pipe frame divided into 4 parts. Also was important to make a choice of tail type and empennage. Many configurations were analyzed and two of them were found to be the best ones: fuselage mounted stabilizers and X-tail empennage. The decision between those two options was based on the fact that torque that will occur takes low value, while landing vulnerability is more meaningful. The fuselage mounted stabilizers was the final choice [8].

6.3 Computional Fluid Dynamics verification

After design original airfoil, it was decided to verify it and do more accurate CFD analysis as shown in Figure 3. Whole aircraft geometry was created in ANSYS software to get as accurate results as possible and to check if the airfoil achieves characteristics like in the vortex lattice method. To find laminar to turbulent transition in boundary layer the Transition SST turbulence model was used. This analysis affirmed previous calculations, so the optimization was correct.

https://doi.org/10.1051/matecconf/201925701003

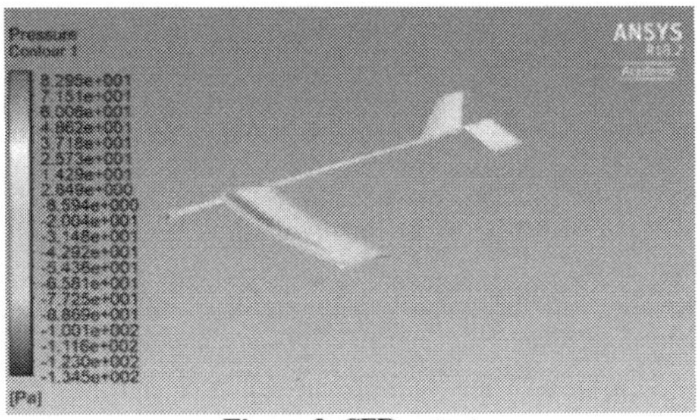

Figure 3. CFD output

6.4 Power Plant Research

One of the most important test was to find the best power plant. The results showed that about 140W electric motor will probably get the best results. The three various motors were selected to perform thorough dynamic thrust tests in wind tunnel, providing for margin of error of algorithm approximations:

- Emax RS2306 2400 RPM/V ~200W
- DRC Inferno Motor 2207.5 - 2550 RPM/V ~200W
- Axi 2208/26 V2 – 1420 RPM/V – 130W

The data collected was concerned, apart from thrust decrement by growth of airflow speed, electric and temperature performance [2]. The final choice was the Axi 2208/26 due to:

- the lowest value of slope of dynamic thrust curve,
- the lowest power consumption that allows to reduce weight by using less capacity battery,
- the lowest temperature in similar conditions – reducing risk of in-flight breakdown

Since the crucial data was collected, it was possible to calculate an exact size of wing and its angle of attack. The second, final optimization software was prepared.

7 Manufacturing and Flight Tests

To make the construction as light as possible it was decided to make it entirely of carbon fiber. Molds were designed and milled for whole aircraft. Wings consist of three, easy to assemble parts and ae made in sandwich technology. Fuselage is seamless, made of halves, glued together under pressure. Stabilizers were made in full core technology using light foam, covered with biaxial carbon fiber with basis weight of 20 grams per square meter. Every element was post cured in temperature 80°C for minimum 15 hours, after initial cure. This process considerably improves strength and heat resistance.

The aircraft ready to fly, without payload, weighs only 278 grams.

As a result of the research the flight tests were performed as shown in Figure 4. They confirmed that developed optimization and manufacturing process was appropriate and done well. All theoretical assumptions were also correct. The airplane stably flied with full assumed payload with dimensions of 10cm x 10cm x 35cm, weighing 1.4 kilograms. The flight tests were done in different weather conditions and payload amount.

Figure 4. Flight tests

8 Conclusions

This paper covers development of geometry and innovative solutions of the light and compact unmanned aerial vehicle, which are optimal for given tasks and meet the initial requirements. Many different ideas were discussed, a compromise between theoretical and practical approaches was reached and the safety boundary was set to be sure that the aircraft can successful fly in any conditions. In addition to that, new tools for aircraft geometry optimization and mass model calculations were developed. To design the best possible UAV many tests, like dynamic thrust of the propulsion system and tensile testing, were done. Innovative manufacturing system was also developed to make as light and durable airplane as possible.

References

1. R. Austin *Unmanned Aircraft Systems* 45-74 (John Wiley and Sons 2010)
2. G. Landolfo *Aerodynamic and structual design of a small nonplanar wing UAV* 6-10, 35-67 (University of Dayton 2008)
3. B. W. McCormick *Aerodynamics aeronautics, and flight mechanics* (John Wiley & Sons, 1979)
4. L. M. Nicolai *Estimating R/C model aerodynamics and performance* **VI** (Lockheed Martin Aeronautical Company 2009)
5. T. C. Witushynsky *Experimental Investigation into the aerodynamics of small air vehicles with* thin *flexible wings,* 16-28 (Case Western Reserve University 2008)
6. T. J. Mueller, J. D. DeLaurier *Aerodynamics of Small Vehicles* **35:** 89-111 (Annual Review of Fluid Mechanics 2003)

7. T. Melin *User's guide and reference manual for Tornado* (KTH 2000)
8. W. Fiszdon *Mechanika lotu* **2** (P.W.N. 1961)
9. A. Skarbinski, W. Stafiej *Projektowanie i konstrukcja szybowcow* (Wydawnictwo Komunikacji i Lacznosci 1966)
10. P. G. Ifju *Flexible-wing-based micro air vehicles* **8** (Mechanical & Aerospace Engineering Department, University of Florida 2005)

Control Oriented Model for Expander Cycle Scramjet

Zhelong Zhao[1], Xianyu Wu[1]

[1] *College of Aerospace Science and Engineering, National University of Defence Technology, Changsha 410073, China*

Abstract. As a efficient and simple design, expander cycle is widely applied in LRE engineering, but it is seldomly used on scramjet research. In order to establish a complete mathematical model for expander cycle scramjet, a control-oriented model for expander cycle scramjet is proposed in this paper. This model consists of four major parts: combustor, cooling channel, turbo pump and nozzle and gives the result of pressure, temperature, mach number and velocity distribution of combustor and cooling channel and is capable of simulate both pure supersonic combustion mode and supersonic shock wave mode of the combustor. Each part is given by specific mathematical description, which contains the calculation of airflow, combustion, heat transfer and thermal cracking of kerosene. By putting all these parts together, a complete model is formed. This model is proposed to calculate the performance and condition of the engine precisely, comprehensively, swiftly and can be directly used in further study.

1 Introduction

Hypersonic airbreathing vehicle has shown its bright future in both military and civil uses. Because of its outstanding performance at the speed range from mach 5 to mach 10, scramjet is thought to be the ideal propulsion system to achieve hypersonic airbreathing flight.

Comparing to other designs, expander cycle is more efficient, more reliable and has a simple structure. The working process of expander cycle scramjet is shown in fig.1. Centrifugal pump pumps cold fuel from the fuel tank to the turbo through cooling channel, within which the cold fuel transfer heat with combustor wall and becomes hot fuel. Hot fuel powers the turbine to drive the centrifugal pump. After flow out of the turbine, the fuel is injected into the combustor to mix with air and release heat after combustion.

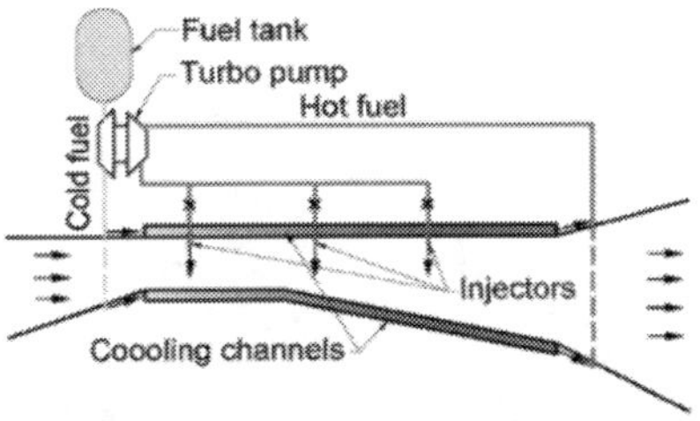

Figure1. Expander cycle scramjet

The design and application of expander cycle fit scramjet perfectly, but there are few research about expander cycle scramjet, and further research requires a model that gives expander cycle scramjet a complete mathematical description. In this paper, aiming on forming a self-holding system that contains the key factors of a kerosene based scramjet, a quasi-one dimensional model for expander cycle scramjet is proposed. This model can be used on analysis the working characteristics and predict the performance of expander cycle scramjet.

2 Modelling

2.1 Modelling of combustor

The condition of the airflow can be defined by its composition, pressure, velocity, temperature, density and mach number. Thus, by setting up and solving the governing equations of the control volume, we can picture the whole flow field of the combustor. The governing equations is set under the following assumptions: a, gas in the combustor is prefect gas which meets the ideal gas equation, b, air flow is steady flow, and the equation does not contain any time term, c, all parameters change only along X axis. Referring to [1], the governing equations is given by:

$$(\rho+d\rho)(A+dA)(u+du)=\rho Au+d\dot{m} \tag{1}$$

$$\frac{1}{u}\frac{du}{dx}=-\frac{1}{\rho u^3}\frac{dp}{dx}-\frac{2C_f}{D}-\frac{(1-\varepsilon)}{\dot{m}}\frac{d\dot{m}}{dx} \tag{2}$$

$$\mathrm{Cp}\frac{dT}{dx}=-\sum_i h_i\frac{dY_i}{dx}-u\frac{du}{dx}+\frac{1}{\dot{m}}\sum_i\left(h_i\frac{d\dot{m}_i}{dx}\right)_{\mathrm{add}}-\frac{2C_f(h_{aw}-h_w)}{DPr^{\frac{2}{3}}}-\frac{h_t}{\dot{m}}\frac{d\dot{m}}{dx} \tag{3}$$

$$\frac{1}{p}\frac{dp}{dx}=\frac{1}{\rho}\frac{d\rho}{dx}+\frac{1}{T}\frac{dT}{dx}-\frac{1}{W}\frac{dW}{dx} \tag{4}$$

$$\frac{dY_i}{dx}=\frac{\dot{\omega}_i W_i A}{\dot{m}}+\frac{1}{\dot{m}}\frac{d\dot{m}_{i,add}}{dx}-\frac{Y_i}{\dot{m}}\frac{d\dot{m}}{dx} \tag{5}$$

$$\frac{dW}{dx}=-W^2\sum_i\frac{1}{W_i}\frac{dY_i}{dX} \tag{6}$$

Equation (1) to equation (6) are the conservation of mass, the conservation of momentum, the conservation of energy, gas equation of state, the conservation of species and the conservation of molecular weight, respectively. The variables of A represents sectional area of the duct, C_f represents skin-friction coefficient of the combustor duct, cp represents specific heat, h represents specific enthalpy, Y represents mass fraction, W represents molecular weight, ε represents the ratio of injection velocity to mainstream velocity. The subscripts of w represents wall, aw represents adiabatic wall, add represents mass addition. The governing equations can be solved in the following form, referring to [8]:

$$\frac{1}{u}\frac{du}{dx}=\frac{1}{(ma^2-1)}\left\{\begin{matrix}\left[\frac{h_t}{C_pT}-kMa^2(1-\varepsilon)-1\right]\frac{1}{q_m}\frac{q_m}{dx}+\frac{1}{A}\frac{dA}{dx}+\frac{1}{W}\frac{dW}{dx}-kMa^2\frac{2C_f}{D}-\\ \frac{1}{C_pT}\frac{1}{q_m}\sum_i\left(h_{t,i}\frac{dq_m}{dx}\right)_{add}+\frac{q_w p_w}{TC_pq_m}+\frac{1}{C_pT}\sum_i h_i\frac{dY_i}{dx}\end{matrix}\right\} \tag{7}$$

$$\frac{1}{\rho}\frac{d\rho}{dx}=\frac{1}{q_m}\frac{dq_m}{dx}-\frac{1}{A}\frac{dA}{dx}-\frac{1}{u}\frac{du}{dx} \tag{8}$$

$$\frac{1}{p}\frac{dp}{dx}=-kMa^2[\frac{1}{u}\frac{du}{dx}+\frac{2C_f}{D}+\frac{(1-\varepsilon)}{q_m}\frac{dq_m}{dx}] \tag{9}$$

$$\frac{1}{T}\frac{dT}{dx}=\frac{1}{p}\frac{dp}{dx}-\frac{1}{\rho}\frac{d\rho}{dx}+\frac{1}{W}\frac{dw}{dx} \tag{10}$$

$$\frac{1}{Ma}\frac{dMa}{dx}=\frac{k}{2}\frac{dW}{Wdx}+\frac{(k-1)}{2C_p}\frac{dC_p}{dx}+\frac{1}{u}\frac{du}{dx}-\frac{1}{2T}\frac{dT}{dx} \tag{11}$$

2.2 Mixing

After the fuel is injected into the combustor, the fuel has to be effectively mixed with air before it can burn. The calculation of mixing efficiency in supersonic flow refers to [2] and is listed as followed:

$$M_c=\frac{u_f-u_a}{a_f+a_a} \tag{12}$$

$$f(M_c)=0.25+0.75e^{-3M_c^2} \tag{13}$$

$$\left(\frac{L_{mix}}{d_f}\right)=\frac{K}{f(M_c)}\left[\frac{\rho_f U_f}{\rho_a U_a}\right]^{\frac{1}{2}} \tag{14}$$

$$\bar{x}=\frac{x-L_{inj}}{L_{mix}} \tag{15}$$

$$\eta_{mix}=1.06492\left(1-e^{-(3.69639\bar{x})^{0.80586}}\right) \tag{16}$$

Where the variables of u_f and u_a represents velocity of fuel injection and airflow, $a_f$and a_a represents speed of sound in the fuel injector and airflow, d_f represents the diameter of fuel injector, L_{inj} represents the axis position of injector, and η_{mix} represents the mixing efficiency. By solving the equations above, we can get the mixing efficiency that increases from none to maximum alongside x-axis.

Even though the method in [2] is raised for the calculation of the mixing between hydrogen and air, but we believe with few modifications of certain parameters, it can also be applied to the calculation of supercritical kerosene mixing.

2.3 Modelling of supersonic shock wave mode

When the equivalence ratio in the combustor reaches a certain level and the pressure in the combustor is high enough to cause boundary layer separation, the scramjet will work under supersonic shock wave mode. In this case, the separation of boundary layer in the combustor will trigger the forming of pre-combustion shock train in the isolator, which compresses the air and change the condition of airflow within the range of the pre-combustion shock train. The boundary layer separation in the combustor will change the shape of actual flow area and change the state of airflow in the combustor. Therefore, the modelling of supersonic shock wave mode is critical.

Through experimental research, the boundary of combustion mode transition is determined in [3]:

$$\left(\frac{M_d}{M_u}\right)_{\text{turbulent}}<0.762 \tag{17}$$

Equation 17 shows that when the ratio of the minimum mach number in the combustor to the mach number at the entrance of the isolator is below 0.762, the scramjet is considered to work under supersonic shock wave mode.

According to [4], the length of pre-combustion shock train can be calculated by:

$$L_{PCST}=\frac{1}{c}ln(\frac{w_2}{w_3}) \tag{18}$$

$$w^2=\frac{(k-1)Ma^2}{2+(k-1)Ma^2} \tag{19}$$

Where the coefficient c can be determined by experiment, [5] suggest that c=0.114.

Based on experimental research, [6] testified the theory that the ratio of the maximum pressure in the combustor to the pressure at the entrance of the combustor is a fixed number when the scramjet is working under supersonic shock wave mode. Which can be described as:

$$p_3=\delta p_{max,comb} \tag{20}$$

Where $p_{max,comb}$ represents the maximum pressure in the combustor and δ is the ratio

Now that we have the values of pressure at the entrance of the isolator, the entrance of the combustor and the maximum pressure in the combustor, we can use polynomial fitting to get the whole pressure distribution, according to [7].

$$\frac{P(x)}{P_2}=1+(\frac{P_3}{P_2}-1)(3-2\chi)\chi^2 \quad (21)$$

$$\chi=\frac{x-x_2}{x_3-x_2} \quad (22)$$

$$\frac{P(x)}{P_3}=1+(\frac{P_t}{P_3}-1)(2-\chi)\chi \quad (23)$$

$$\chi=\frac{x-x_3}{x_4-x_3} \quad (24)$$

Equation 21 and equation 23 are the polynomial fitting equations to solve the pressure distribution in the pre-combustion shock train and in the combustor under the influence of the separation of boundary layer, respectively.

3 Cooling channel

Cooling channels are placed around the combustor to absorb the heat from the combustor wall. Regenerative cooling can keep the combustor wall from over temperature while giving the fuel a certain amount of heat sink.

3.1. Heat transfer

The heat transfer between the cooling channel and the combustor wall includes both convective heat transfer and radiant heat transfer.

3.1.1 Convective heat transfer

According to [3], the heat flux density of convective heat transfer can be calculated in the following form:

$$r=\sqrt[3]{Pr^*}=\sqrt[3]{\frac{\mu^* C_p^*}{\lambda^*}} \quad (25)$$

$$h^*=\frac{h+h_w}{2}+0.22r\frac{u^2}{2} \quad (26)$$

$$Re_x=\frac{puwx}{RT^*\mu^*} \quad (27)$$

$$q_c=\frac{0.0287puw(h_{aw}-h_w)}{RT^* Pr^{*\frac{2}{3}} Re_x^{*\frac{1}{5}}} \quad (28)$$

Where μ represents viscosity, q_c represents the heat flux density of convective heat. Variables with superscript '*' represent they are calculated under reference temperature. [3] suggest that $Pr^*=0.9$. The skin-friction coefficient of the combustor duct can be calculated with the reference tempurature solved.

$$C_f=\frac{0.0574}{Re_x^{*\frac{1}{5}}}\frac{T}{T^*} \quad (29)$$

3.1.2 Radiant heat transfer

The radiant heat transfer between gas and combustor wall can be calculated as:

$$q_r=\sigma\varepsilon_{w,ef}\left(\varepsilon_g T_{g,s}^{\ 4}-a_w T_{g,w}^{\ 4}\right) \quad (31)$$

Where σ represents Boltzmann's constant, $\varepsilon_{w,ef}$ represents effective blackness, ε_g represents theblackness of gas, $T_{g,s}$ represents the temperature of gas, $T_{g,w}$ represents the temperature of inner combustor wall.

The total heat flux density is:

$$q_w=q_r+q_c \quad (32)$$

3.2 Flow in channel

The basic idea of solving the flow identity in the cooling channel is the same as the solving in combustor. Here we directly put forward the solving equations, referring to [9].

$$\frac{du_f}{dx}=(-\frac{A_c p_f q_w r_{cw}}{T_f C_{pf} q_{mf}}+p_f\frac{dA_c}{dx}-\frac{1}{2}\frac{f}{D_{hc}}q_{mf}u_f)/(q_{mf}-\frac{A_c p_f}{u_f}-\frac{A_c p_f u_f}{T_f C_{pf}}) \quad (33)$$

$$\frac{dT_f}{dx}=\left(\frac{q_w P_w}{q_{mf}}-u_f\frac{du_f}{dx}\right)/C_{pf} \quad (34)$$

$$\frac{dp_f}{dx}=p_f(\frac{1}{T_f}\frac{dT_f}{dx}-\frac{1}{u_f}\frac{du_f}{dx}-\frac{1}{A_c}\frac{dA_c}{dx}) \quad (35)$$

$$\frac{d\rho_f}{dx}=\rho_f(\frac{1}{\rho_f}\frac{dp_f}{dx}-\frac{1}{T_f}\frac{dT_f}{dx}) \quad (36)$$

Where r_{cw} represents the wetted perimeter of combustor wall, q_{mf} represents the mass flow of the coolant fuel. Variables with the subsript 'f' represtet the sates of the coolant fuel. The temperature of the sombustor wall can be sloved useing basic heat conduct theory.

3.3 Thermal cracking

For the long chain hydrocarbon, when the temperature raises to a certain level, thermal cracking will happen and the long chain hydrocarbon will crack into hydrocarbon with shorter carbon chain. The temperature and pressure in the cooling channel is often beyond the cracking limit, and the thermal property of the coolant fuel changes severely due to thermal cracking. Therefore, the process of thermal cracking must be included in the modelling of hydrocarbon fuel scramjet. According to [9], the process of light thermal cracking of decane can be predicted by PDD model :

$$\begin{aligned} C_{10}H_{22}\rightarrow & 0.151H_2+0.143CH_4+0.256C_2H_4 \\ & +0.126C_2H_6+0.23C_3H_6+0.18C_3H_8 \\ & +0.196C_4H_8+0.102C_4H_{10}+0.171C_5H_{10} \\ & +0.124C_5H_{12}+0.195C_6H_{12}+0.089C_6H_{14} \\ & +0.169C_7H_{14}+0.152C_8H_{16}+0.012C_8H_{18} \\ & +0.053C_9H_{18}+0.003C_9H_{20} \end{aligned}$$

The reaction rate constant can be calculated as:

$$k_A=1.6\times10^{15}\times exp(-\frac{263.7}{RT}) \quad (37)$$

With the thermal cracking model and the calculation method of thermal cracking rate, we can generally predict the thermal cracking process of decane.

4 Turbo pump

Turbo pump is a critical part of expander cycle scramjet. It uses the extra heat from the combustor and charges the fuel. Turbo pump is an important part to keep scramjet work properly.

4.1 Centrifugal pump

The torque of centrifugal pump can be calculated as:

$$M=\frac{dL}{dt}=\rho q(R_2c_{p2}-R_1c_{p1}) \quad (38)$$

Where M represents torque, L represents moment of momentum, R represents the radium, c represents the velosity. Variables with the subsript 'p1' represtet entrance of the pump, 'p2' represents the exit of the pump.

The lift of the pump can be calculated as:

$$H=H_m\frac{n^2}{n_m^2}=\left(k_2q_{Mm}^{\ 2}+k_1q_{Mm}^{\ 2}+k_0\right)\frac{n^2}{n_M^2}$$
$$=k_2q_m^{\ 2}+k_1q_m\frac{n}{n_M}+k_0\frac{n^2}{n_M^2} \quad (39)$$

Where q_m represents mass flow, n represents rotary speed, n_M represents reference rotary speed. And k_0, k_1, k_2 are paramters which can be solved by experiment or numerical calculation. For a given centrifugal pump, its lift can be directy calculated using equation 39. Based on the design of turbo pump in [10], k_0=5.256, k_1=1.16, k_2=-0.625, n_M=25000, q_m=1.

4.2 Turbine

Consider the flow of the fuel in the turbine as an adiabatic expansion process:

$$\left(\frac{Pt_1}{P_{t0}}\right)^{(k-1)/k}=T_{t1}/T_{t0} \quad (40)$$

Where the variables with the subscript 't0' represents before the static blade, 't1' represents after the static blade and before the rotating blade, 't2' represents after the rotating blade. The conservation of energy in the static blade of the turbine can be described by the conservation of stagnation enthalpy:

$$i_o^{\ *}=i_1^{\ *} \quad (41)$$

$$i_0^{\ *}=c_pT_0^{\ *}=c_pT_0+C_0^{\ 2}/2 \quad (42)$$

$$i_1^{\ *}=c_pT_1^{\ *}=c_pT_1+C_1^{\ 2}/2 \quad (43)$$

$$\frac{c_1^{\ 2}-c_0^{\ 2}}{2}=c_p(T_0-T_1) \quad (44)$$

Where the variables with the superscript '*' represents stagnation. The isentropic expansion work of the turbine is:

$$E_{ad}=i_{t0}^{\ *}-i_{t2}^{\ *}=Cp(T_{t0}^{\ *}-T_{t2ad}^{\ *}) \quad (45)$$

$$\frac{T_{t0}^{\ *}}{T_{t2ad}^{\ *}}=\left(\frac{P_{t0}^{\ *}}{P_{t2}^{\ *}}\right)^{\frac{k-1}{k}} \quad (46)$$

$$E_{ad}=CpT_{t0}^{\ *}\left(1-\frac{1}{(\pi_p^{\ *})^{\frac{k-1}{k}}}\right)=E_{ad}^{\ s}-c_{t2}^{\ 2}/2 \quad (47)$$

$$P_T=\eta_T\dot{m}_TE_{ad}^{\ s} \quad (48)$$

$$M_T=P_T/\omega \quad (49)$$

Where E_{ad} represent the isentropic expansion work, $E_{ad}^{\ s}$ represents the ideal isentropic expansion work, M_T represents the torque of the turbine, ω represents the rotary speed.

4.3 Turbo pump assembly

As the turbine and the centrifugal pump are rigidity connected, and we ignore all time related terms, the transmission relationship between them is simple:

$$\omega_T=\omega_P \quad (50)$$

$$M_P=M_T \quad (51)$$

5 Result and conclusion

5.1 Result

We test the model by setting the boundary conditions with the mach number of 2.7, the temperature of 1160K, the pressure of 0.1MPa. Assuming the pressure at the exit of the turbo to be 1Mpa, the pressure in the fuel tank to be 0.5Mpa, the fuel temperature to be 300K. The geometry structure of the scramjet is:

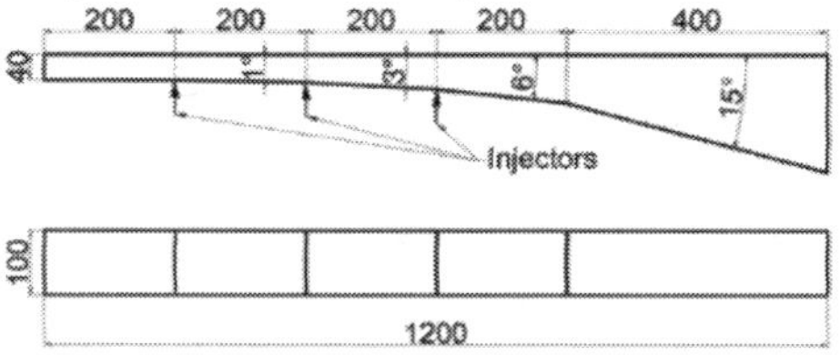

Figure 2. Geometry structure of the combustor.

The first 200mm is the isolator, followed by the combustor, which consists of three 200mm-long-expanding sections. The expansion angle is 1 degree, 3 degree and 6 degree, respectively. Three fuel injectors are placed at the very front of the three expanding sections. At the end is the nozzle, which is 400mm long and has a expansion angle of 15 degree.

5.1.1 Example 1

In the first example, all three fuel injectors inject fuel at the same mass flow with the total equivalence ratio of 0.6. At this state, the scramjet should be working under supersonic shock wave mode, according to equation 17. We calculate the case using both pure supersonic combustion mode and supersonic shock wave mode, and the result is represented in blue and red line respectively, as is shown in Fig.3. By comparing the results, the influence of the pre-combustion shock train can be seen.

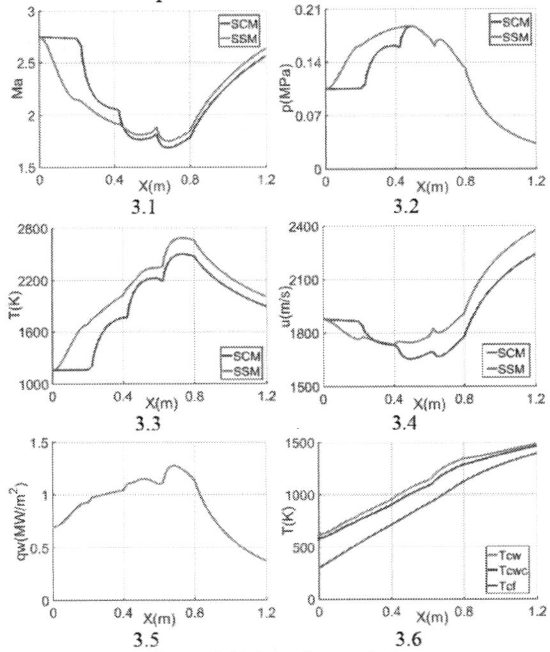

Figure3. Result of example 1.

Fig 3.1 to Fig 3.5 are the distributions of mach number, pressure, temperature, velocity, heat flux, respectively. Fig 3.6 shows the temperature distribution of combustor wall and coolant fuel, Tcw represents the inner side of combustor wall, Tcwc represents the outer side of combustor wall, Tcf represents the coolant fuel. Fig 3.5 and Fig 3.6 are calculated under shock wave mode.

According to the result of example 1, the model can correctly predict the influence of fuel injection, fuel combustion, heat transfer and the geometric change of the duct to the airflow. And the calculation of supersonic shock wave mode gives an expected result of the combustor flow field under the influence of pre-combustion shock train and boundary layer separation. The result shows that the temperature and velocity in supersonic shock wave mode is relatively higher than in pure supersonic combustion mode.

5.1.3 Example 2

The second example is to test how the rotating speed of turbo pump react to the increasing total equivalence ratio. Like example1, all fuel injectors inject fuel at the same mass flow. The result is shown in fig.4.

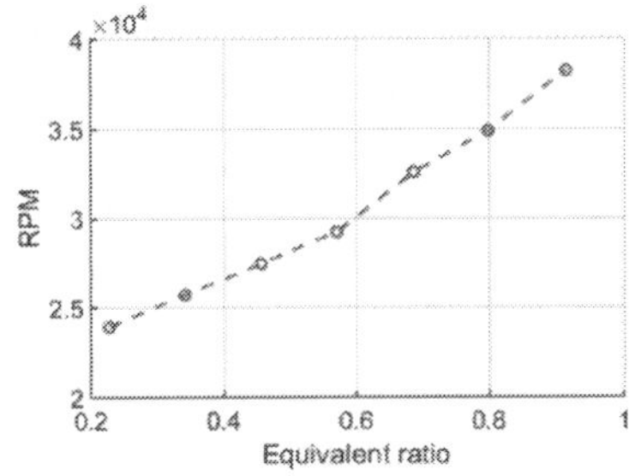

Figure4. Result of example 2.

The rotary speed of turbo pump goes up with the increase of the total equivalence ratio, and the increasing rate of rotary speed increases after the scramjet entries supersonic shock wave mode.

5.2 Conclusion

A model of a kerosene based expander cycle scramjet is proposed. The simulation of the examples gives an expected result, and it have demonstrated that the model can predict the working condition of the expander cycle scramjet swiftly and correctly. With a few throttle devices added, it can be directly used in the simulation of the controlling process of the expander cycle scramjet. Moreover, by adding models of certain parts or processes to the model, it can be further modified. In the next step of work, the result of the model should be compared to experiment or numerical simulation and adjustment of parameters should be made to improve the accuracy of the model.

References

1. O'Brien, T. F., Starkey, R. P., and Lewis, M. J., "Quasi-One-Dimensional High-Speed Engine Model with Finite Rate Chemistry,"Journal of Propulsion and Power, Vol. 17, No. 6, 2001, pp. 1366–1374.
2. Birzer, C., and Doolan, C. J., "Quasi-One-Dimensional Model of Hydrogen-Fueled Scramjet Combustors," Journal of Propulsion and Power, Vol. 25, No. 6, 2009, pp. 1220–1225.
3. Heiser, W., and Pratt, D., *Hypersonic Airbreathing Propulsion*, AIAA,Washington, D.C., 1994, pp. 479–485.
4. Torrez S M, Dalle D J, Driscoll J F. New Method for Computing Performance of Choked Reacting Flows and Ram-to-Scram Transition[J]. Journal of Propulsion and Power, 2013, 29(2): 433-445.
5. Ikui T, Matsuo K, Nagai M. The Mechanism of Pseudo-Shock Waves[J].Bulletin of JSME, 1974, 17(108): 731-739.
6. Zhang Peng, Yu Gang, The study of one-dimensional flow analysis model of the combustor in supersonic combustion experiments [J]. Experiment and Measurements In Fluid Mechanics, 2003 (01): 88-92.
7. Billig F S, Dugger G L. Predictions of Precombustion Wall Pressure Distributions in Scramjet Engines [J]. Journal of Spacecraft and Rockets, 1973(9): 620-622.

8. Zhang D, Feng Y, Zhang S, et al. Quasi-One-Dimensional Model of Scramjet Combustor Coupled with Regenerative Cooling[J]. Journal of Propulsion & Power, 2016, 32(3):1-11.
9. WARD T, ZABARNICK S, ERVIN J, et al. Simulations of flowing mildly-cracked normal alkanes incorporating proportional product distributions[J]. Journal of Propulsion & Power, 2004, 20(3): 394-402.
10. Xuan Jin. Study on Operation Characteristics of Scramjet with Expansion Cycle. [D]. Changsha: National University of Defence Technology, 2016

The Effects of Non-uniform Distribution of Oxidizer Flow on High-Frequency Combustion Instability

Peng Chen [1], Wansheng Nie [1], Kangkang Guo [1], Xing Sun [1], Yu Liu [1], Haiqing Wang [1]

[1] *Space Engineering University, 101416 Beijing, China*

Abstract. The numerical calculation of three-dimensional unsteady combustion for the combustion chamber of LOX/kerosene high pressure staged combustion rocket engine was carried out. By changing the offset ratio of oxygen mass flow rate in the edge area of the injector face, computational studies were conducted to investigate the effects of non-uniform distribution of oxidizer flow on combustion instability for a liquid-propellant rocket engine. The calculation results show that the offset ratio of oxygen mass flow rate changes the distribution of heat release in the combustion chamber. Within a certain range of offset ratio, the non-uniform distribution degree of oxidizer flow enhances the coupling between the pressure and heat release. As a result, it leads to an increase in the pressure oscillation amplitude in the combustion chamber. However, if the offset ratio is too large, the oxygen-fuel ratio will be too small in some regions, which will reduce coupling between the pressure and heat release and increase the damping of combustion instability.

1 Introduction

Almost every rocket engine in the development process encounters the problem of combustion instability. In particular, high-frequency combustion instability could destroy the cooling structure of the combustion chamber, and instantly ablate the chamber wall, resulting in destroying the structure of combustion chamber. High frequency combustion instability in liquid propellant rocket engines is widely believed to result from a coupling of the complex and dynamic processes of injection, atomization, vaporization, mixing, and chemical reaction with a characteristic response of the gas dynamics in the combustion chamber [1]. Understanding and predicting high-frequency combustion instability in liquid-propellant rocket engine continues to pose significant challenges due to the highly complex and nonlinear nature of turbulent combustion processes. The F-1 rocket engine in the United States has experienced multiple combustion instability during the development process. After conducting more than 2,000 full-scale experiments, it has well suppressed combustion instability [2]. But so far, there is no comprehensive model that can accurately predict the degree of combustion instability in certain engines and certain operating conditions [3].

The cost of conducting full-scale engine experiments is high, and due to the high temperature and pressure environment inside the rocket engine, the data measured in the experimental study is limited. Therefore, Computational Fluid Dynamics (CFD) is needed to analyse the flow field and chemical reaction in the rocket engine in more details and obtain the data of the full flow field [4]. Various studies have been conducted using CFD to study combustion instability. Urbano et al. have explored a multiple-injector LOX/GH_2 rocket engine by making use of Large Eddy Simulations (LES) [5]. A single-element GO_2/GH_2 shear-coaxial injector was studied by Masquelet et al. [6] And Morgan et al. studied transverse combustion instability by changing injector configurations in a rectangular chamber rocket engine which uses RP-1, ethane and oxygen as propellants [7].

The LOX/kerosene high pressure staged combustion rocket engine has the advantages of high reliability, environmental protection, low cost and high performance. It is of great practical significance to analyse the combustion instability of the LOX/kerosene staged combustion cycle rocket engine. In this paper, computational studies were conducted to investigate the effects of non-uniform distribution of oxygen mass flow on combustion instability for the LOX/kerosene high pressure staged combustion rocket engine.

2 Calculation model

A model was established for the combustion chamber of the LOX/kerosene rocket engine. Oxygen is a gas phase and is treated as a continuum. Kerosene is a liquid phase and its spray process uses the Discrete Phase Model (DPM) which based on the Euler-Lagrange approach. The combustion model uses the Finite-Rate/Eddy-Dissipation model with one-step reaction mechanism. The calculation uses the Reynolds-Averaged Navier-Stokes (RANS) modelling and pressure implicit splitting of operator (PISO) algorithm to simulate the unsteady compressible flow under the three-dimensional condition [8].

2.1 Physical model

The injector distribution of the injector face is shown in the Figure 1. High-pressure oxygen-enriched gas is injected from the center hole of the gas/liquid coaxial swirling injector and kerosene is injected from the swirl hole on the sidewall. The effects of non-uniform distribution of oxygen mass flow on combustion instability is studied by changing the offset ratio of oxygen mass flow rate.

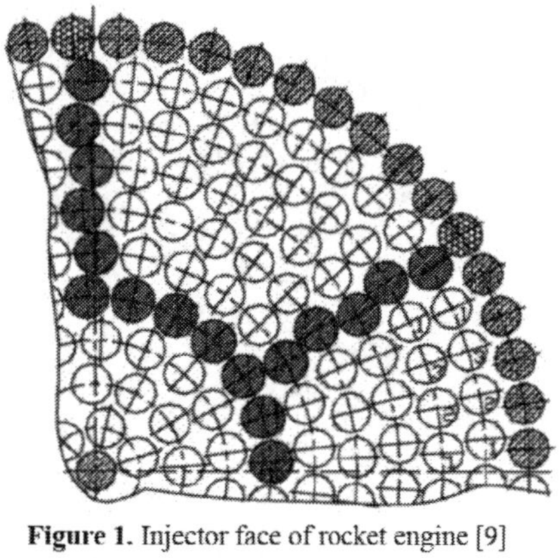

Figure 1. Injector face of rocket engine [9]

2.2 Gas governing equations

Numerical calculations are performed in 3D full size. The general format of the control equation in the three-dimensional rectangular coordinate system is as follows:

$$\frac{\partial(\rho\phi)}{\partial t}+div(\rho\vec{U}\phi)=div[\Gamma_\phi grad(\phi)]+S_\phi+S_\phi^C+S_\phi^L \quad (1)$$

Where the Γ_ϕ is generalized diffusion coefficient and the is S_ϕ generalized source item. S_ϕ^C is source item of chemical reaction. S_ϕ^L is source item of liquid phase evaporation. Then the continuity equation, momentum conservation equation and energy conservation equation can be derived from this equation by making ϕ and Γ_ϕ take different parameters.

The most widely used RANS equations is currently used. Further, the standard *k–ε* turbulence model is Robustness, economy, and reasonable accuracy for a wide range of turbulent flows, so this article uses the standard *k–ε* turbulence model.

2.3 Discrete Phase Model

The Euler-Lagrange approach with DPM is used where the gas phase is treated as a continuum, while the dispersed liquid phase is solved by tracking a large number of particles through the calculated flow field, i.e. computing the particle trajectories at specified intervals during the fluid phase calculation [10].

2.4 Finite-Rate/Eddy-Dissipation turbulent combustion model

The net reaction rate of Finite-Rate/Eddy-Dissipation was controlled by Arrhenius rate and Eddy-Dissipation reaction rate, taking the smaller of the two rates.

2.5 Chemical Reaction Mechanism

The chemical reaction uses a one-step reaction mechanism:

$$C_{12}H_{23}+17.75O_2\rightarrow 12CO_2+11.5H_2O$$

2.6 Mesh and boundary conditions

As shown in the figure 2, the combustion chamber is divided into triangular prism grids with the element number of 510,000. Oxygen is a gas phase, and the DPM model is used for the spray process of kerosene. The spray angle is 40°, and the swirl fraction is 0.5. The droplet diameter uses the Rosin-Rammler distribution function with an average diameter of 2.5 μm. The inlet boundary condition is mass-flow-inlet, and the outlet boundary condition is pressure-outlet. The walls are assumed to be adiabatic and are treated as no-slip boundaries.

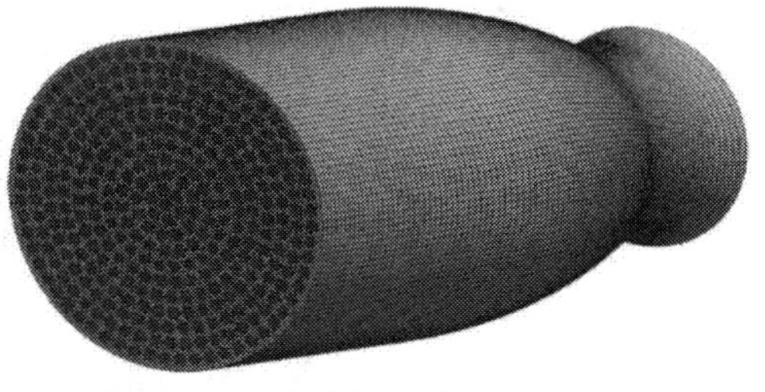

Figure 2. Mesh of the combustion chamber

3 Calculation

3.1 Model validation

The test pressure of the combustion chamber was 18 MPa. And in figure 3, the numerical calculation shows that the mean pressure in the combustion chamber is 17.6 MPa with an error of 2.2%, which validate the reliability of the numerical simulation model built by this paper.

https://doi.org/10.1051/matecconf/201925701005

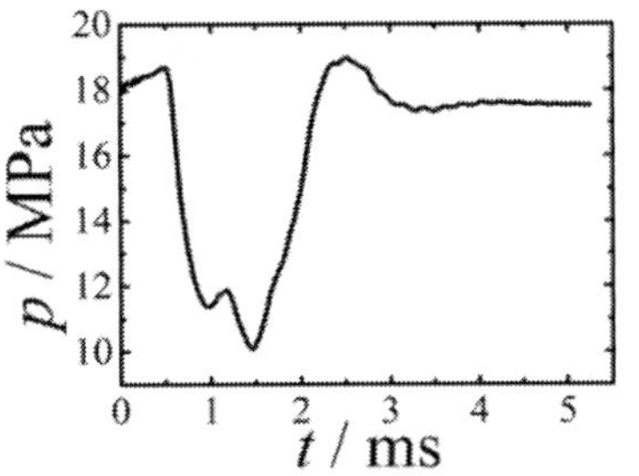

Figure 3. Pressure of the combustion chamber

3.2 Oxygen mass flow offset

The oxygen mass flow rate was 371 kg/s, and the kerosene mass flow rate was 60 kg/s. As a result, the oxygen-fuel ratio was 6.2. The total mass flow rate of oxygen in the outermost two cycles of the injectors remains stable. It is divided into two half-ring areas A and B. Then set the oxygen mass flow in these areas to a certain proportion to achieve the purpose of flow offset. So case1-case3 are set up separately, and let the A, B area oxygen mass flow ratio be 3.5:1, 5:1, 8:1. In case0, mass flow rate of oxygen in outermost two cycles is uniform, which is the reference for other cases. Four monitor points point1-4 are set on the cross-section 10 mm from the face as shown in figure.

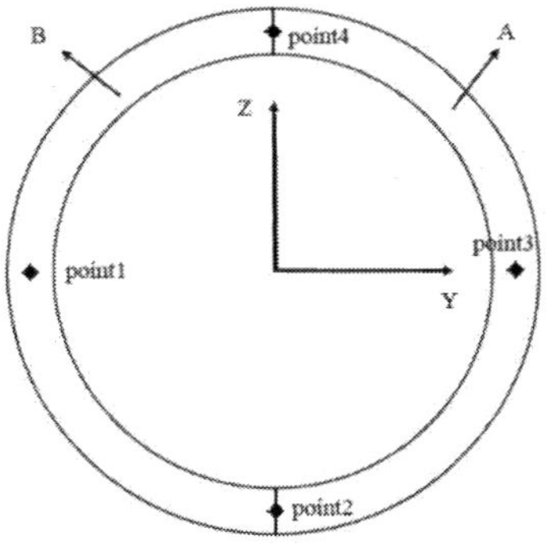

Figure 4. The area divided and the monitor points

3.3 Results

The pressure history at the monitor points of case 0-3 is shown in the figure. It can be seen from the figure that there are pressure oscillations at all monitor points. It is considered that the combustion instability occurs when the pressure oscillation amplitude in the combustion chamber exceeds 10% of the average pressure [11]. As shown in the Table 1, the maximum pressure oscillation amplitude of case1 and case2 exceeds 20% of the average pressure, and combustion instability is obvious; while the maximum pressure oscillation amplitude of case0 is close to 10%, and combustion instability is not obvious; case 3 shows no combustion instability.

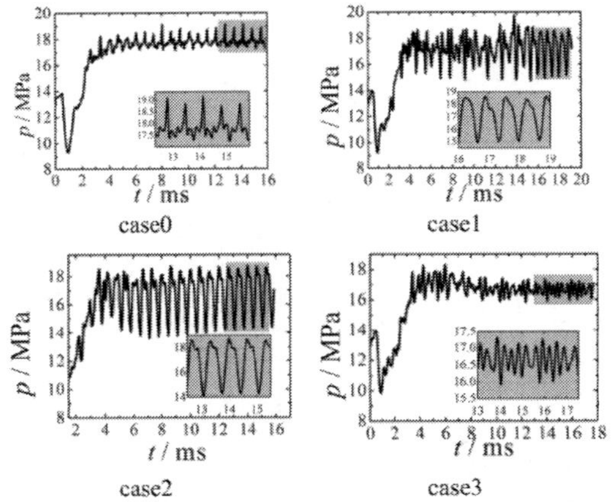

Figure 5. Pressure of combustion chamber in the four cases

Table 1. the result of the four cases

	case0	case1	case2	case3
Offset ratio	1:1	3.5:1	5:1	8:1
Mean pressure /MPa	17.8	17.3	17.2	16.9
$p'/\bar{p}$	10.8%	22.8%	27.2%	7.9%
Oxygen-fuel ratio of area B	6.2	3.4	2.5	1.9

The pressure frequency of the four monitor points was analyzed using the Fast Fourier Transform (FFT). The dominant frequency of case1 and case2 is 1500 Hz, and the frequency is greater than 1000 Hz, indicating that there was high-frequency combustion instability in the combustion chamber; the main frequency of case0 is 1462Hz, but the amplitude of main frequency is obviously lower than that of case1 and case2; the main frequency of case3 is 2867Hz, and the amplitude is the smallest among the four cases.

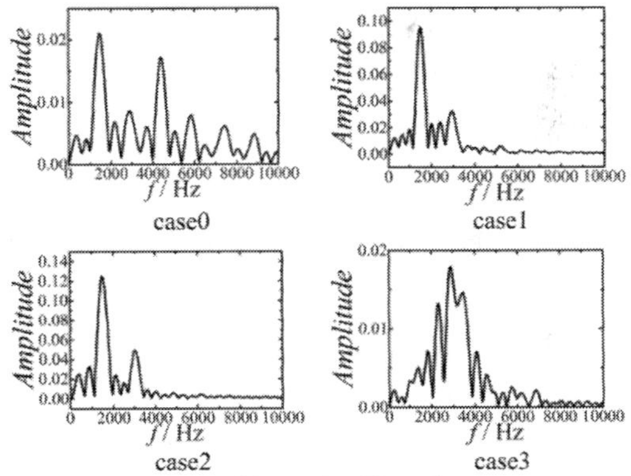

Figure 6. FFT result

The first-order tangential frequency of the combustion chamber is 1661 Hz. And the error with case1 and case2 is about 9.7%. Then the pressure contours of the cross-section x=10 mm from the face within one cycle

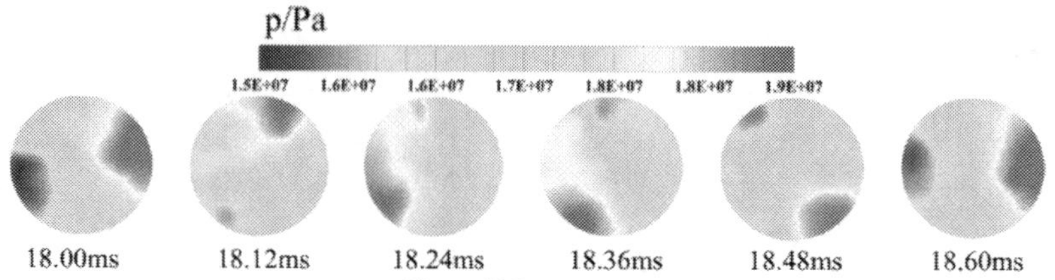

Figure 7. Pressure contours

is shown in the Figure 7, it can be found that there is a rotating high-pressure region and low-pressure region in the combustion chamber, which is the first-order tangential mode of rotation.

4 Analyse

Rayleigh criterion is often mentioned when combustion instability is studied [12]. It shows that if heat is added to a pressure oscillation at maximum pressure or if heat is extracted at minimum pressure the pressure oscillation amplitude will increase. In contrast, if the heat is added at minimum pressure or extracted at maximum pressure, the amplitude will decrease [12,13]. In the combustion chamber, due to the processes of injection, atomization, vaporization and mixing may have a certain disturbance to the flow field, and the combustion process is also very sensitive to changes in the flow field parameters, there will be a random oscillation of heat release rate [5]. Further, heat release oscillations can cause pressure oscillations. If the heat release oscillation frequency and the pressure oscillation frequency are coupled with the inherent acoustic frequency of the combustion chamber, the pressure oscillation in the combustion chamber will continue to propagate and amplify, eventually causing combustion instability.

The pressure and heat of reaction of point4 in case1 and case2 are shown in figure 8. It can be clearly seen that when the pressure rises, the heat release rate of the chemical reaction also increases at the same time. Both the frequency and the phase are basically same, indicating that there is a strong coupling relationship between the heat release rate and the pressure.

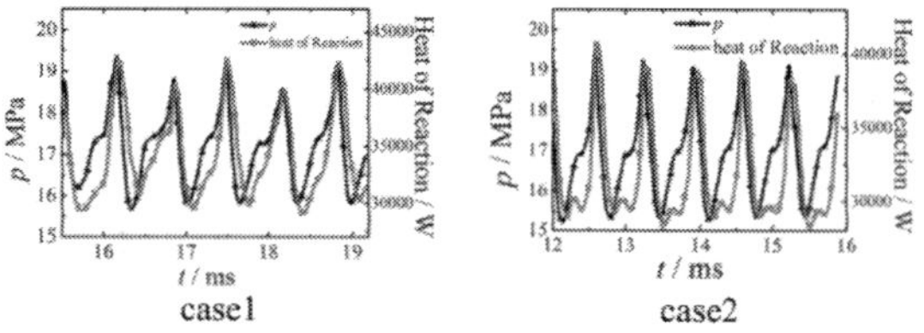

Figure 8. Pressure and heat of reaction in case1 and case2

Under different conditions of offset radio, there are different amplitudes of pressure oscillations, indicating that non-uniform distribution of oxygen mass flow has a great influence on the pressure oscillation amplitude of the combustion chamber. In the process of increasing the offset ratio of oxygen from a completely uniform increase to 3.5:1 to 5:1, the pressure oscillation amplitude in the combustion chamber increased from 10.8% to 22.8% to 27.2%. It can be clearly seen that non-uniform distribution of oxygen mass flow can drive combustion instability.

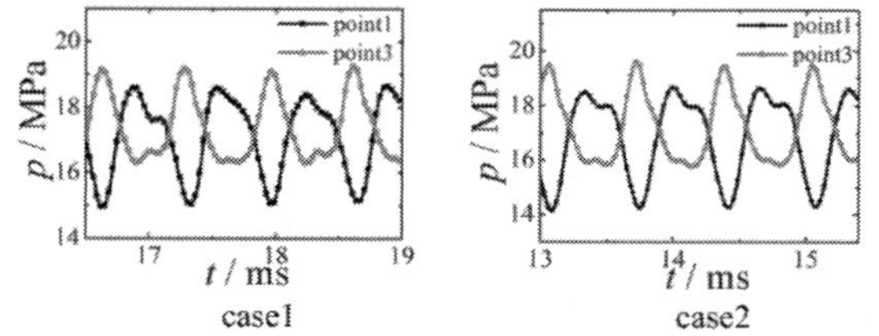

Figure 9. Pressure for point1 and point3

As shown in the figure 9, the pressure for point1 and point3 of case1 and case2 are plotted where point1 is in the B area and point3 is in the A area. During the rotation of the high pressure region and the low pressure region in areas A and B, the pressure peaks at the two monitor points are not consistent. It can be seen that the maximum pressure in the high pressure region increases from area B to area A, and decreases from area A to area B. And The minimum pressure in the low pressure region also increases from area B to area A, and decreases from area A to area B. Point 1 and point 3 have the same frequency of pressure oscillation and opposite phases, which indicats that when the high pressure region rotates to area A, the low pressure region rotates exactly to area B. Due to the large oxygen-fuel ratio in area A, kerosene complete combustion results in greater heat release. In contrast, the oxygen-fuel ratio in area B is relatively small, so that kerosene is not completely combusted resulting in a small heat release radio, as shown in Figure 10. In this situation, the degree of coupling between the non-uniform heat release and the pressure is further aggravated. According to the Rayleigh criterion, the coupling drives the combustion instability. On the other hand, when the high pressure region rotates to area B and the low pressure region rotates to area A, the maximum pressure in the high pressure region has a certain attenuation, and the minimum pressure in the low pressure region has a certain increase. However, they can always maintain a certain level until they enter the next cycle, so that pressure oscillations can be maintained and developed.

Finally, due to the dissipation of the wall and other factors in the combustion chamber, a pressure oscillation cycle is formed when the dissipation effect and the coupling of pressure and heat release are balanced. Within this offset radio range, the greater the offset ratio, the stronger the effect of heat release on the pressure oscillation and the larger the pressure oscillation is.

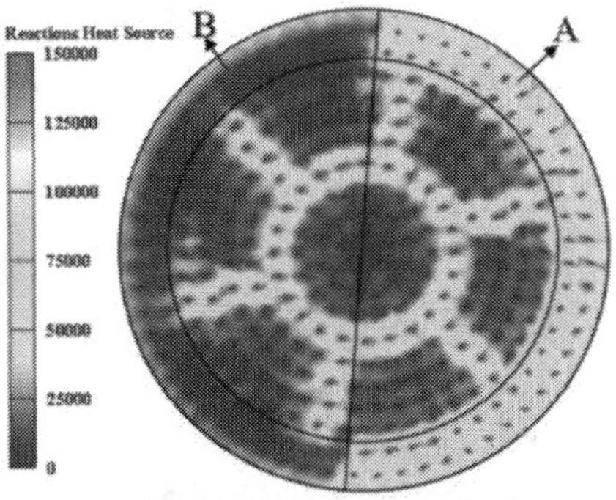

Figure 10. Heat release cloud

As the offset ratio increased from 5:1 to 8:1, the oxygen-fuel ratio of the area B decreased from 2.5:1 to 1.9:1, which is far below the oxygen-fuel ratio of 3.4:1 for the propellant just to complete combustion. Therefore, the area B is not fully burned, and the heat release is small, which has a great impediment to the pressure oscillation, resulting in no complete pressure oscillation cycle in the combustion chamber, so no combustion instability occurs.

Therefore, there should be an appropriate offset ratio of oxygen mass flow that maximizes amplitude of pressure oscillation.

5 Conclusion

In this paper, the effect of non-uniform distribution of oxygen mass flow on combustion instability for a liquid-propellant rocket engine is investigated by changing the offset ratio of oxygen mass flow rate in the edge area of the injector face. Within a certain range, with the increase of the offset ratio of oxygen mass flow, the pressure oscillation peak also increases, which indicates that the non-uniform distribution degree of oxidizer flow enhances the coupling between the pressure and heat release. It can also be seen that there should be an appropriate offset ratio that can maximize the amplitude of pressure oscillation when combustion instability occurs in the combustion chamber of LOX/kerosene rocket engine. In this regard, during the design of the rocket engine, the propellant flow distribution should be arranged properly to reduce the probability of combustion instability during the injection process when the propellant is just injected.

References

1. K. Miller, J. Sisco, N. Nugent, W. Anderson, Journal of Propulsion and Power, 23(5),1102(2007)
2. J.C. Oefelein,V. Yang, Journal of Propulsion and Power, 9(5), 657(1993)
3. V. Sankaran, M. Harvazinski , W. Anderson, D. Talley, ICCFD,7, 3105(2012)
4. R. Smith, G. Xia, W. Anderson, Combustion Theory and Modelling, **16(2)**, 341 (2012)
5. A. Urbano, L. Selle , G. Staffelbach , B. Cuenot, T. Schmitt, S. Ducruix, S. Candel, Combustion and Flame, **169**,129 (2016)
6. M.Masquelet, S.Menon, Journal of Propulsion and Power,**26.5**,924(2010)
7. C.Morgan, K.Shipley,W.Anderson, Journal of Propulsion and Power, **31(6)**,1696(2015)
8. K. Guo, W. Nie, Y. Liu, H. Cai, ICMAE2017,719 (2017)
9. L. Li, J. Chen, Z. Liu, Missiles and Space Vehicles,**313**,16 (2011)
10. Kozic, S. Ristic, M. Puharic, B. Katavic, Third Serbian Congress on Theoretical and Applied Mechanics,290 (2011)
11. Y.Ga, H.Zhang, X.Wang, Joural of Engineering Thermophysics,**33.5**,883,(2012)
12. M.Harvazinski,W.Andersony,C.Merkle, Aiaa/asme/sae/asee Joint Propulsion Conference & Exhibit,47(2011)
13. L.Rayleigh, The Theory of Sound, **2** (1896)

The Research of Autonomous In-orbit Health Management System for China's Manned Spacecrafts

Linhou Bai[1], Hong Yang[1] and Ke Liang[1]

[1]*China Academy of Space Technology, Manned Space Department, 100094 Beijing, China*

Abstract: The development trend of manned spacecraft is long-term in-orbit flight and long-term astronaut's residency. Under the condition of limited TT&C resource, the autonomous health management of spacecraft is a critical and mandatory system to ensure the safety of the astronauts and platform. Based on the discussion of the autonomous health management objectives and management modes of manned spacecraft and combining the distribution and hierarchical information structure of China's manned spacecraft, this paper elaborates an autonomous health management system. This autonomous health management system adopts a by-level safety threshold management method by establishing a mapping database between different important risk safety thresholds and system safety modes. When the key parameters reach different pre-set thresholds, the system autonomously downgrades and reconstructs the aircraft functionalities in the event of failure to realize optimal auto-processing and solves the issue between randomness of auto-processing and task continuity. The research has been applied to the China's Space Station.

1 PREFACE

From the beginning of the 21st century, the systematic health management of aircraft has emerged as a specialized technology in aviation and aerospace fields. Representative methods include Complex System Integrated Health Management (CSIHM), Prognostics and Health Management (PHM), Integrated Vehicle Health Management (IVHM), and Integrated System Health Management (ISHM) [1-2].

Domestic research on aircraft health management has been carried out lately, relevant researches mainly focus on spacecraft fault diagnosis management, autonomous operation management, and integrated health management of complex systems. Some subsystems in such as Resource 21 satellite, Resource 22 satellite and ShenZhou spacecraft have adopted the intelligent technologies such as fault isolation, diagnosis, and system reconstruction, which have indicated good performance.

The development trend of manned spacecraft is long-term in-orbit flight and long-term astronaut's residency. Under the condition of limited TT&C resource, the autonomous health management of spacecraft is a critical and mandatory system to ensure the safety of the astronauts and platform[3]. Based on the information architecture of manned spacecraft in China, this paper elaborates an autonomous health management system which by adopting a by-level safety threshold management method achieves spacecraft functional degrade and reconstruction in the event of failure.

2 MANNED SPACECRAFT HEALTH MANAGEMENT

The manned spacecraft health management comprehensively evaluates the spacecraft's health status through the telemetry data and images, and predicts the performance changes of the spacecraft, so as to execute the detection, isolation and recovery of the emergent faults, and to prevent possible failures. The aim is to ensure the spacecraft normal operation to meet the mission target. It generally includes two parts: spacecraft autonomous health management and ground health management.

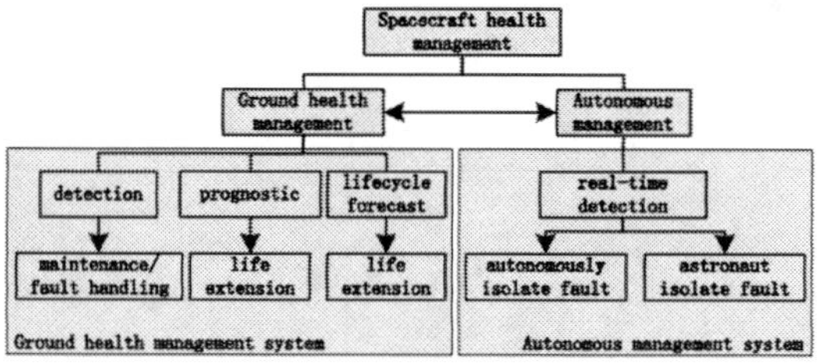

Figure 1. Spacecraft health management structure

(1) Ground health management can use spacecraft telemetry, image, and astronaut voice to evaluate the system health status using ground-based flight control expert system, perform ground action to

anomaly, or predict possible anomaly with preventive measures implemented. The shortcomings of ground health management are TT&C resources of land-bases, ocean-bases and relays cannot be occupied exclusively by manned spacecraft for a long period of time, so the ground cannot receive real-time spacecraft data over a long period of time.

The ground health management system evaluates the health status of the entire spacecraft, the reliability of sub-units and system, and in-orbit status to make lifecycle forecast. The evaluation provides decision support for spacecraft replenishment, maintenance and life extension, and generally contains the following functionalities:

1) Store telemetry data and ground test data and perform big data processing on massive data. By using various intelligent inference algorithms (such as physical model, neural network, data integration, fuzzy logic, expert system, etc.) [4] to extract reliability data, and to extract and save lifecycle prediction data.
2) Import the extracted reliability data into the assessment SW for reliability assessment.
3) By collecting actual in-orbit flight data and ground life test data, the life-time trend forecast curve is modified and fitted, to achieve in-orbit life-time prediction.

(2) The autonomous management system can monitor and sense its own status, with autonomously detecting, isolating, and recovering from failures[5-7]. With consideration of computer processing capability, its main management scope includes platform safety and health status which requires high real-time processing capability. Meanwhile with astronaut's presence, the astronauts can observe and perceive the real state of the spacecraft, and take proper operation, which is also within the scope of autonomous management. In other words, the health monitoring, diagnosis, and processing performed by manned space vehicles or completed by astronauts are all within the scope of in-orbit autonomous health management. The shortcomings of autonomous health management is that it cannot judge and process all the conditions. In particular, when there is a fault outside the plan, it can only be resolved by the ground health management system. The autonomous health management is mainly used to isolate major emergency faults, while the ground system executes the final solution.

The characteristics of ground health management and autonomous health management are as follows:

Table 1. Ground and autonomous health management Comparison

Spacecraft health management	**Advantage**	**Disadvantage**
Ground health management	Ability to achieve comprehensive management of spacecraft status and life-time prediction	Cannot receive real-time spacecraft data over a long period of time
Autonomous health management	Ability to detect the health status of spacecraft and autonomously isolate faults	only manage the key health status of spacecraft, and no ability to predict lifecycle

3 MANNED SPACECRAFT AUTONOMOUS HEALTH MANAGEMENT GOALS

The manned spacecraft autonomous health management is mainly designed to achieve the following goals:

(1) Secure the safety of astronaut and platform: for the faults that affect the safety of astronauts or cause platform failure, the health management can promptly discover, respond and dispose, to ensure the safety of astronauts and platforms, and gain time for ground handling. In the event of an emergent major failure, it provides timely response and disposal, so that the aircraft can quickly enter the safety mode for energy, attitude, propulsion, and thermal control accordingly. It alerts astronaut and displays the fault information, that the astronauts can timely proceed based on the fault information and according to the prepared fault solution.

(2) Support long-term operation: In the long-term operation stage, the TT&C resources are limited. The autonomous health management system ensures timely detection and diagnosis outside the TT&C area to carry out disposal in real time. Within TT&C area, the fault information can be received by the ground for further analysis and disposal.

(3) Release partial work of flight control on the ground: Transition from traditional ground-monitoring mode relying on people, to autonomous health management design that autonomously detect and diagnose the failures to reduce misjudgment and missed judgment by people. It releases partial work of flight control on the ground and hence reduces operating costs.

4 AUTONOMOUS HEALTH MANAGEMENT MODE

A good design of autonomous health management system can greatly enhance the spacecraft's in-orbit independency and safety, and hence reduce the need for TT&C resources. For example, the China's Tiangong-1 target aircraft has autonomous management design for energy module, in-orbit attitude control module and thermal control module[8]. During the long-term in-orbit period, the TT&C resources demand has been reduced from 80% to 10% compared with the rendezvous and docking tasks, which has increased the efficiency of ground-side monitoring and reduced in-orbit operating costs.

Considering the computer capability and design complexity, normally the system mainly includes the platform security related status that requires high real-time performance into the scope of autonomous management. The China's Space Station has numerous failure modes in-orbit, with variety of types and complex state. At the space station assembly phase, the number of nodes plus visiting aircrafts can reach to 7 at the most, and there are crossing connections of information, power supply and control, which causes more system coupling, and more complex coupling faults. In order to accurately define the management scope of such large-scale spacecraft autonomous management systems, FMEA (Failure Mode and Effect Analysis) and FTA (Fault Tree Analysis) are normally used jointly, combining the static functional modules of the spacecraft and dynamic flight mission profile to ensure comprehensive and accurate identification. Generally, the modes in which the manned spacecrafts require autonomous health management are shown in the following table:

Table 2. Manned spacecraft autonomous health management mode

No.	Autonomous Management Mode	Fault Analysis	Fault Criteria
1	Power supply capability reducing	Battery pack failure Solar panels stuck	Battery discharged greater than expected
2	Power bus off-duty	Power supply equipment failure Load short-circuited	Power bus voltage drops 1553B bus communication abnormal
3	Heat dissipation capability reducing	Radiator leakage	Device temperature rising Fluid circuit pressure dropping
4	Heat dissipation off-duty	Radiator leakage Failure of fluid circuit pump causing malfunction of the fluid circuit	Device temperature rising Fluid circuit pressure dropping Abnormal pump speed Abnormal pressure difference between inlet and outlet
5	Capsule pressure dropping	Capsule broken by micro-meteor Capsule seal broken	Capsule pressure dropping Temperature of some equipment dropping abnormally
6	Capsule misfire	Load short circuited	Fire detection sensor alarm Camera image display
7	Abnormal attitude control	Attitude sensor malfunctional Propulsion system malfunctional Propellant leakage	Poor attitude Abnormal status of attitude sensor Propulsion pipe pressure dropping
8	Unable to establish space-earth link	Communication device malfunctional Abnormal attitude	Abnormal status of communication device Poor attitude

5 AUTONOMOUS HEALTH MANAGEMENT SYSTEM ARCHITECTURE

The main functionalities of the autonomous health management system include spacecraft health status acquisition, health management data transmission, health management data storage, fault detection, fault diagnosis and failure decision.

Achieving autonomous health management functionalities generally requires equipment such as Central processor, Data acquisition equipment, Commanding equipment, Data storage and Functional terminals[9-11]. The Central processor receives the data from Data acquisition device and judge whether the fault occurs. At the same time, it requires the Commanding equipment to send order to the Functional terminals to perform the fault isolation operation. It controls Data storage to record the fault condition. The Data acquisition equipment receives the aircraft status data and sends the data to the central processor for data analysis. The functional terminal receives and executes the fault handling instructions from Commanding equipment to make sure that the spacecraft is in safe mode in orbit. By scheduling from the Central processor, the Data storage equipment stores the status data collected by the Data acquisition equipment as well as the autonomous health management processing and result data when outside of the flight control area, so to support later fault analysis by ground expert.

Figure 2 is a block diagram of spacecraft autonomous health management system.

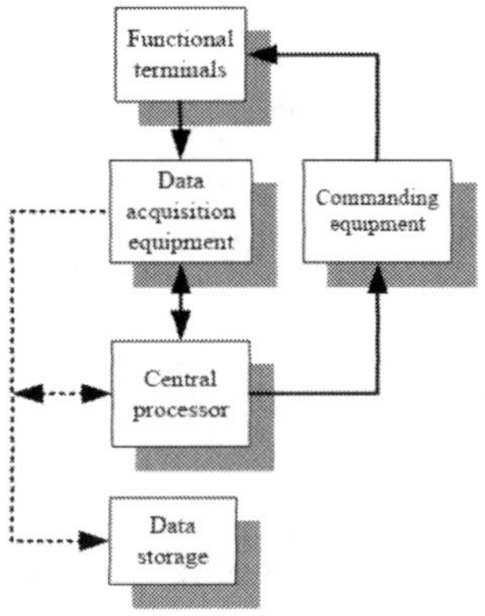

Figure 2. Spacecraft autonomous health management system block diagram

Regarding the realization of manned spacecraft autonomous health management, taking into account that the manned spacecraft information system is a distributed embedded system, in order to increase the response speed to failures, reduce the core processing computing burden, a by-level health management system architecture is adopted. The autonomous health management functionalities are completed by numerous computers scattered in different subsystems. Because each computer has different level in the information system, the tasks it undertakes for health management are not the same, and are generally divided into three levels. The top layer is the core processing unit, the second layer is the control computer or regional controller of each subsystem, and the third layer is the subsystem internal subnet equipment and other terminal.

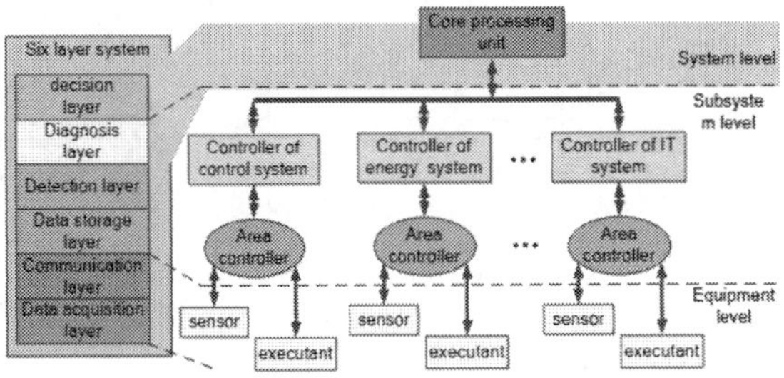

Figure 3. Mapping between Autonomous Health Management function block and info-system by-level hierarchy model

6 OPERATING MODE

The autonomous health management system in-orbit repeatedly detects the health status of the basic functional systems, including the information system, energy system, attitude & orbit control system, human environment control system and cabin structure detection system. For major emergency failures, the space station directly enters the safe mode for emergency disposal, and then waits for further investigation and disposal from the ground side. For normal failures, it provides alarm according to established strategies, or performs fault handling such as cutting off and resetting. The autonomous health management system informs the astronaut through image, alarm and other means of detected platform health status information, fault diagnosis and disposal information and simultaneously send to the ground.

With testability design, the autonomous health management system collects the status information step by step and layer by layer and performs fault detection and diagnosis.

(1) Equipment level

Through the testable design such as device's BIT, HW monitoring circuit and test point, the health management system obtains the device status information, and carry out fault detection and diagnosis, and report the fault information to the subsystem controller. For devices that are directly connected to the main bus of the system network, such as digital control device and communication control device, the fault information is directly reported to the core processing unit.

(2) Subsystem level

The subsystem controller collects the health status of the equipment in the subsystem. According to the configured fault mode and programs of detection & diagnosis, it executes fault detection and diagnosis and report the fault information to the core processing unit.

If the subsystem controller judges the security mode to be triggered, it will perform emergency handling, and at the same time notify the core processing unit. Or it will directly notify the core processing unit and leave to core processing unit to implement the emergency handling.

(3) System level

The core processing unit integrates fault information of the related subsystems to determine whether the safe mode entry condition is fulfilled and perform emergency processing if it's fulfilled. All the fault information is sent to ground base as well as to panel subsystem to display. The astronauts carry out further actions based on the pre-established plan.

7 SECURITY THRESHOLD MANAGEMENT AND TROUBLESHOOTING

An one-for-all disposal will disable some of the functionalities of the aircraft necessarily, so a by-level safety threshold management method is adopted, to establish a mapping database with different safety threshold of critical risks and system safety modes. When critical parameters reach different preset thresholds, the spacecraft functionalities are turned-off level by level, which are sending alarms, shutting down high power devices, and closing of payloads. With the spacecraft functional degrade and reconstruction, the autonomous disposal result is optimized, which solves the problem between randomness of auto-processing and task continuity. The specific analysis process is as follows:

(1) Follow the method described in chapter 3 to identify the safety related fault mode.
(2) Classify the disposals of different fault modes and determine the corresponding system safety mode after the disposal. Minimize the processing steps as few as possible to have one disposal solution to cope with multiple fault modes, so to enhance the convenience of implementation and improve the system reliability. For example, the disposal for both power supply capacity reducing and heat dissipation capability reducing is to shut down devices. Therefore, one shut-down device method could be considered to handle both scenarios.
(3) Map the spacecraft functional degrade levels to different levels of safety mode: According to the severity of fault, the spacecraft functionalities are degraded and mapped to the corresponding level of

safety modes, so repeat the down-grade and mapping that eventually a minimum safety mode is formed. This mode contains functionalities that ensure the autonomous flight of spacecraft. The minimum safety mode requires the spacecraft functionalities including downlink engineering telemetry data, ground command receiving and data importing, thermal environment assurance and jet-mode attitude control.

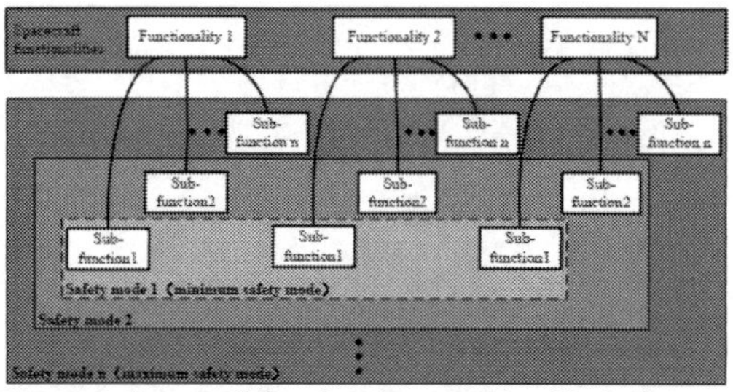

Figure 4. Spacecraft function decomposition block diagram

(4) Verify procedure if down-graded to minimum safety mode: Design a procedure to set the spacecraft to the minimum safety mode, which is programmed in blocks according to different safety modes and sets relevant important functionalities to a relatively backward position in the procedure. When a fault occurs, the functionalities in the safety mode relating to the fault are disabled, and the lower level safety mode is entered. Level by level disable functionalities and enter next lower level safety mode, until the minimum safety mode is entered.

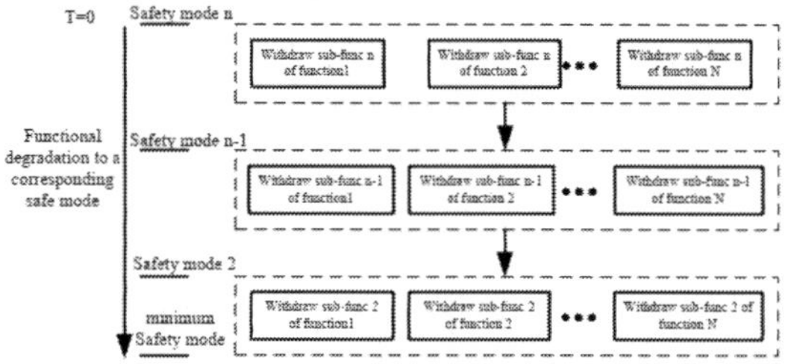

Figure 5. Spacecraft functional degradation block diagram

Take a case of spacecraft power supply capacity decreasing as an example, the mapping between different safety thresholds and system safety modes is shown in the following table:

Table 3. Mapping of safety thresholds and system safety modes in the case of spacecraft power supply capability decreasing

Power supply decreasing	Safety mode	Recovery method
Battery discharged more than expected	Kept functions: high and low speed telemetry, remote control, thermal controlled environment, attitude accurate control (control moment gyro), payload function	Alarm to astronaut and ground base
Battery discharged more than 50%	Kept functions: High and low speed telemetry, remote control, thermal controlled environment, attitude accurate control (control moment gyro)	Withdraw payload function: turn off the load device
Battery discharged more than 60%	Kept functions: High and low speed telemetry, remote control, thermal controlled environment, attitude general control (Jet control)	Withdraw attitude accurate control function: turn off the control moment gyro
Battery discharged more than 70%	Kept functions: Low speed telemetry, remote control, thermal controlled environment, attitude general control (Jet control)	Withdraw high-speed data transmission: turn off power-consuming devices such as relay KSA terminals
Unibus off-duty	Kept functions: Low speed telemetry, remote control, thermal controlled environment, attitude general control (Jet control)	Withdraw high-speed data transmission: turn off power-consuming devices such as relay KSA terminals, and switch the function to another bus for power supply

(5) The spacecraft automatically performs functional degradation to a corresponding safety mode: the spacecraft decides faults according to preset threshold, and when a fault occurs, the spacecraft automatically performs functional degradation processing, following the pattern of by-level safety mode n down to the minimum safety mode. Data is real-time collected in the process of disposal to determine whether the fault disappears and set the spacecraft to the corresponding safety mode. If the fault disappears, the current functional degradation processing module is closed, and the subsequent processing program is stopped. Figure 6 shows the specific disposal process.

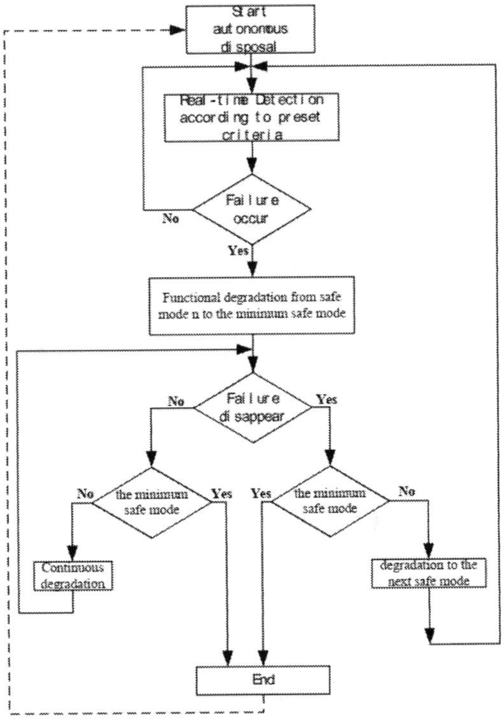

Figure 6. Safety mode disposal flowchart

(6) Ground-side fault diagnosis and treatment: The spacecraft is set and maintains in safety mode after fault occurs. When spacecraft enters the monitoring and control zone, the ground diagnoses and disposes faults based on the analysis of data recorded in the spacecraft data storage and from the real-time telemetry.

8 CONCLUSIONS

In this paper, based on the information architecture of China's manned spacecraft, an autonomous health management system is proposed and a by-level safety threshold management method is adopted to achieve spacecraft functional degrade and reconstruction in the event of failure to realize optimal auto-processing.

References

1. Zeng Sheng-Kui, Michael G P, WU Ji. Acta Aeronautica ET Astronautica Sinica, **26**, 626—632(2006)
2. Zhu Ye, Cheng Kai. Technology Wind, **5**, 59(2012)
3. Zhu Yi-Lin. Spacecraft Engineering, **18**, 13-20(2009)
4. Sun Bo, Kang Rui, Xie Jin-Song. Systems Engineering and Electronics, **10**, 1762-1767(2007)
5. Liang Ke, Deng Kai-wen, Ding Rui, et al. MANNED SPACEFLIGHT, **20**, 116~121(2014)
6. Felke T, Hadden G, Miller D, et al. Proc. 2010 IEEE Aerospace Conference, Atlanta,Georgia: 20-22(2010)
7. Benedettini O,Baines T S, Lightfoot H W, et al. Proceedings of the Institution of Mechanical Engineers, Part G: Journal of Aerospace Engineering, **223**, 157-170(2009)
8. Zhi-yong. Spacecraft Environment Engineering, **28**, 525-528(2011)
9. Li Qing, Sun Guo-Jiang, Li Xiao-Tong. Spacecraft Environment Engineering, **29**, 574-578(2012)
10. Li Qing-dong, Zhang Qing-zhen, Ren Zhang, et al. CONTROL TECHNOLOGY OF TACTICAL MISSILE, **55**, 107-110(2006)
11. Yuan Qin-cheng, Zhang Zhong-hua, Wu Chuan-hong. ELECTRICAL EQUIPMENT, **2**, 41-44(2001)

Grid Convergence Analysis for MUSCL-based Numerical Scheme in Shockwave-containing Flows

Guangning Li[1,2], Dinesh Bhatia[3], Min Xu[1] and Jian Wang[2]

[1]*School of Astronautics, Northwestern Polytechnical University, Xi'an, Shaanxi, 710072, China*
[2]*School of Mechanical and Aerospace Engineering, Faculty of Science, Engineering, and Computing, Kingston University, Roehampton Vale campus, Friars Avenue, SW15, 3DW, London, UK*
[3]*Military Technological College, Muscat Governorate, Oman*

Abstract. This paper investigated the influence of limiter functions widely utilized in MUSCL-type (Monotone Upstream-centred Schemes for Conservation Laws) upwind numerical schemes on the solution accuracy of shockwave-containing flows. An incident shock interacting with laminar boundary layer developed on a flat plate was numerically simulated with the in-house developed code. A mixed-order grid convergence study was performed to assess the spatial errors of different limiters in simulating the selected shockwave-containing flow on flat plate. The conclusions are that, limiter functions implemented in the current in-house code play the critical roles in accurately predicting shockwave-containing flows. The mixed-order error estimator based on grid convergence study was proved to be applicable to evaluate the spatial errors of shockwave-containing flows, where the shock could reduce the nominal second- or third-order accuracy to first-order. The mixed-order estimator is conservative in the sense that the actual error is less than the error estimated, in the examined case.

1 Introduction

As the advances in computing power, there are great demands for the high confidence in modeling and simulating engineering problems. It has become a trend to employ high-order shock capturing schemes with at least a third-order spatial accuracy to perform investigations for shockwave-containing flows[1–3]. However, the traditional second-order upwind schemes are more preferable in practice due to their simplicity, flexibility and robustness. For strong discontinuity-containing flows there are two different strategies to conduct second-order accurate schemes: MUSCL type and non-MUSCL type. Van Leer's MUSCL approach[4] is the widely accepted method to achieve a second-order upwind scheme, and the MUSCL interpolation process can be enhanced with the so-called limiter functions or limiters to prevent the overshoot or undershoot in numerical solutions if discontinuous solutions existing in the flow field. The purpose of utilising a limiter function is to suppress the non-physical oscillations and maintain monotonicity while retaining the expected second-order accuracy in numerical solutions.

For shockwave containing flows, the most common numerical methods are characteristic-based upwind methods that can achieve nominally third-order accuracy at most. However, the solution accuracy can reduce to first order at discontinuities due to the presence of shock waves. Previous work by Roy et al.[5-7] verified that, the use of the nominal second-order scheme in discontinuity-containing flows can result in the presence of both first- and second-order error in solutions. The non-monotonic grid convergence behaviour was found to occur when the first- and second-order error terms were of opposite signs, thus leading to error cancellation. When the solution variables behave in a non-monotonic fashion, a mixed-order error estimator by Roy[5] could be applied to assess the spatial errors of mixed-order solutions. This mixed-order error estimator was proven to provide conservative error estimate in the sense that the actual error is less than the estimated mixed-order error. Therefore, Roy's mixed-order error estimator is employed here to verify the solution accuracy of the selected limiter functions.

The limiter functions used in conducting second-order upwind schemes have great influence on the numerical resolutions of expansion fans, contact discontinuities and shock waves in flow fields. A brief description of the numerical method and the in-house code used in current study are provided In Section 2, and the formulation of MUSCL interpolation with limiter functions are also presented in details. A mixed-order error analysis method is introduced in Section 3 to prepare for grid-independent study and to investigate the influence of limiters on the accuracy of the solution of shockwave-containing flows. The influences of various limiters on accuracy, and grid convergence behaviours of numerical are demonstrated in Section 4. Finally, meaningful conclusions on the properties of the widely accepted limiters in shockwave-containing flows are drawn.

2 Numerical Method

The two-dimensional compressible Navier-Stokes equations with thin shear layer approximation in the generalized coordinate (ξ,η) plane are given in conservation form as follows:

$$\frac{\partial Q}{\partial t}+\frac{\partial (F-F_v)}{\partial \xi}+\frac{\partial (G-G_v)}{\partial \eta}=0 \quad (1)$$

Where Q is conservative vector, F and G are convective flux vectors, F_v and G_v are viscous flux vectors.

An in-house CFD code named ATACF (Analysis Toolkits for All speed Compressible Flows), which is the updated version of ATTF (Analysis Toolkits for Transonic Flows) in [8], developed by Li, is applied to perform the simulation in the current research. In ATACF, the convective terms have several options to be discretized: central or upwind-biased methods, including central-difference scheme, flux-vector splitting scheme (FVS), flux-difference splitting schemes (FDS) and the series of AUSM (Advection Upstream Splitting Method) schemes. As compared with FVS and FDS schemes, the AUSMpw+ scheme[9] tends to be more attractive as an improvement of the AUSM scheme[10]. Therefore, the AUSMpw+ scheme is employed to perform the simulations to verify the properties of the relevant limiter functions in the present paper.

2.1 Limiter Functions

A brief review of the original MUSCL interpolation equations proposed by van Leer[4] is presented as follows:

$$(q_L)_{i+\frac{1}{2}}=q_i+\frac{1}{4}[(1-\kappa)\Delta_-+(1+\kappa)\Delta_+]_i \quad (2)$$
$$(q_R)_{i+\frac{1}{2}}=q_{i+1}-\frac{1}{4}[(1-\kappa)\Delta_+ +(1+\kappa)\Delta_-]_{i+1}$$

Where q_L and q_R are the approximate solutions of state variables at left and right sides of the cell interfaces, and the primitive variables $q=(\rho\ u\ v\ p)^{\mathrm{T}}$ are used to obtain the interpolated state values. $\Delta_+=q_{i+1}-q_i$ is forward difference operator and $\Delta_-=q_i-q_{i-1}$ is backward difference operator. As noted by Anderson[11], by switching from $\kappa=-1$ to $\kappa=1/3$ the accuracy of smooth solutions can be increased, while $\kappa=0$ and $\kappa=1/3$ are the most popular options in practice.

To suppress these unphysical oscillations in solutions, monotonicity-preserving principles are invoked in MUSCL interpolation stage by introducing limiter functions. Following the treatment of Spekreijse[12], the interpolation formula shown in Eq.(2) is rewritten as:

$$(q_L)_{i+\frac{1}{2}}=q_i+\frac{1}{2}\psi(r_i)(\Delta_-)_i \quad (3)$$
$$(q_R)_{i+\frac{1}{2}}=q_{i+1}-\frac{1}{2}\psi(1/r_{i+1})(\Delta_+)_{i+1}$$

Where $r=\Delta_+/\Delta_-$, and $\psi(r)$ is the limiter function described as follows:

$$\psi(r)=\frac{1}{2}[(1-\kappa)+(1+\kappa)r]\phi(r) \quad (4)$$

Four commonly accepted limiter functions within the framework of second-order limiter region are briefly reviewed as following:

Van Leer limiter:

$$\psi(r)=\frac{r+|r|}{1+r} \quad (5a)$$

Van Albada limiter:

$$\psi(r)=\frac{r^2+r}{1+r^2} \quad (5b)$$

Hemker-Koren limiter:

$$\psi(r)=\frac{2r^2+r}{2r^2-r+2} \quad (5c)$$

Min-mod limiter:

$$\psi(r)=\frac{1}{2}[(1-\kappa)\phi_1(r)+(1+\kappa)\phi_2(r)] \quad (5d)$$

Where

$$\phi_1(r)=\max[\min(1,\beta r),0],\quad \phi_2(r)=\max[\min(r,\beta),0],\quad 1\le\beta\le\frac{3-\kappa}{1-\kappa}$$

With different κ and β, the Min-mod limiter in term of Eq.(5d) has different forms and functionalities:

1) with $\beta=1$, Eq. (5d) becomes as following:

$$\psi(r)=\max[\min(r,1),0]=\begin{cases}1 & r\ge 1\\ r & 0\le r\le 1\\ 0 & else\end{cases} \quad (6a)$$

2) with $\kappa=0$, $\beta=3$

$$\psi(r)=\begin{cases}2r & 0\le r\le 1/3\\ (1+r)/2 & 1/3\le r\le 3\\ 2 & r\ge 3\\ 0 & else\end{cases} \quad (6b)$$

3) with $\kappa=1/3$, $\beta=4$

$$\psi(r)=\begin{cases}2r & 0\le r\le 1/4\\ (1+2r)/3 & 1/4\le r\le 4\\ 3 & r\ge 4\\ 0 & else\end{cases} \quad (6c)$$

3 Grid Convergence Analysis Method

The mixed-order extrapolation method by Roy[5] is presented as in Eq.(7), where three discrete solutions on three different grids from fine to coarse are needed:

$$\begin{aligned}f_1&=f_{exact}+g_1h_1+g_2h_1^2+O(h_1^3)\\ f_2&=f_{exact}+g_1h_2+g_2h_2^2+O(h_2^3)\\ f_3&=f_{exact}+g_1h_3+g_2h_3^2+O(h_3^3)\end{aligned} \quad (7)$$

Where f_k is a discrete resultant value on grid level k with "$k=1$" indicating the finest grid. f_{exact} is the exact solution, g_i is the i th-order error term coefficient and h_k stands for the measure of grid spacing on the level k. Without loss of generality, by defining the finest grid spacing as unity (i.e. $h_1 = 1$), the grid refinement factor is defined as

$$r_{k,k+1} = h_{k+1} / h_k \quad (8)$$

The above Eq.(7) may be solved approximately for g_1, g_2 and f_{exact} with constant grid refinement factor r as follows:

$$\tilde{g}_1 = \frac{r^2\varepsilon_{21} - \varepsilon_{32}}{r(r-1)^2} \quad (9)$$

$$\tilde{g}_2 = \frac{\varepsilon_{32} - r\varepsilon_{21}}{r(r+1)(r-1)^2} \quad (10)$$

$$\tilde{f}_{exact} = f_1 + \frac{\varepsilon_{32} - (r^2+r-1)\varepsilon_{21}}{(r+1)(r-1)^2} \quad (11)$$

Where $\varepsilon_{32} = f_3 - f_2$ and $\varepsilon_{21} = f_2 - f_1$.

In general, the above estimation of the solution $\tilde{f}_{exact}$ is of third-order accuracy, and the estimated spatial error based on Eq.(11) on each grid level is calculated using the following relationship:

$$|spatial\ \text{error}\ (\%)| = \left|\frac{f_k - \tilde{f}_{exact}}{\tilde{f}_{exact}}\right| \times 100 \quad (12)$$

As stated by Roy [5], if multiplied by a safety factor, for example $F_s = 3.0$, Eq.(12) will produce the similar error estimate with that of the GCI (Grid Convergence Index) method proposed by Roache [13,14] when the generalized Richardson Extrapolation method is employed to determine $\tilde{f}_{exact}$ in Eq.(12) and when the estimated GCI error is less than 10-20%. Thus Eq.(12) is employed in current research to evaluate the spatial error of the grid convergence study.

The normalized magnitudes of the first- and second-order error terms, as well as their sum, are formulated as:

$$\left|\frac{\tilde{g}_1 h}{\tilde{f}_{exact}}\right| \times 100 \quad (13.a)$$

$$\left|\frac{\tilde{g}_2 h^2}{\tilde{f}_{exact}}\right| \times 100 \quad (13.b)$$

$$\left|\frac{\tilde{g}_1 h + \tilde{g}_2 h^2}{\tilde{f}_{exact}}\right| \times 100 \quad (13.c)$$

4 Results and Discussions

An incident shock interacting with laminar boundary layer developed on a flat plate, as shown in Fig.1 is studied. This case study corresponds to the experiments by Hakkinen et. al. in [15], which is often used to validate the efficiency and accuracy of numerical methods.

Three computational grids are generated in a rectangle domain with length $2L$ starting from the leading edge of the plate and height L starting from the surface of the plate. The reference length measured from the leading edge of the plate to the impingement point on the plate is L. The detailed grid information used for this computation is shown in Table.1. The free stream conditions are given as: $Ma_\infty = 2.0$, $T_\infty = 117K$ and the Reynolds number based on the reference length is $Re_\infty = 2.96\times10^5$, the incident shock angle is $\beta = 32.6°$. Table.1 shows the detailed grid information used in current simulations including grid dimensions used.

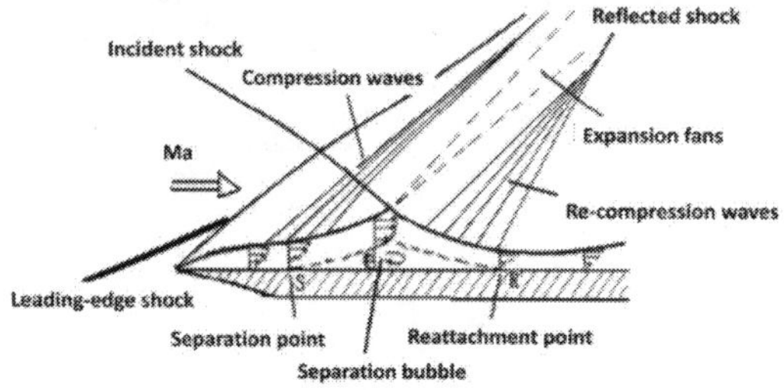

Figure 1: Schematic of shock-wave interactions with laminar boundary layer on the flat plate.

Table 1: Detailed information of Grids

Grid number	Grid points	Grid spacing h
1	401×161	1
2	201×81	2
3	101×41	4

Fig.2 shows the predicted drag coefficients, as well as the extrapolated grid-independent drag coefficients with different limiters. It is evident that, as the grid is successively refined, the numerically predicted drag coefficients for each limiter converged to their grid-independent values respectively. Fig.2 also shows that the Min-mod limiter with $\beta = 4$ predicted the highest grid-independent drag coefficient, while the Hemker-Koren limiter predicts the lowest drag coefficient.

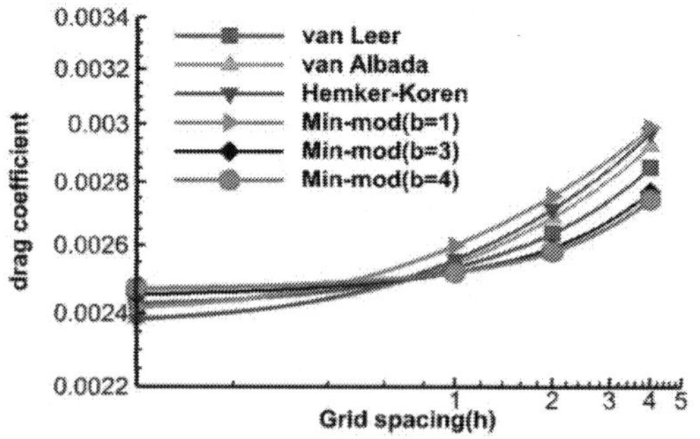

Figure 2: Drag coefficients on different grids for incident shock on the flat plate.

In Fig.3 the grid convergence errors of first- and second-order terms and their sum from Eq.(13), as well as the discrete solutions from Eq.(12) are plotted together. Only three limiters of van Albada limiter, Hemker-Koren limiter and Min-mod limiter with $\beta = 1$ show the non-monotonic convergence behaviour due to the cancellation of first- and second-order terms with opposite signs. However, the grid-convergence errors for all limiters approach to formally second-order range asymptotically on coarser grids, while to observed first-order range on finer grids. The spatial errors of the solution with the employment of limiter functions in the numerical scheme are of the mixed-order accuracy in current simulation. Fig.4 shows the spatial error levels with different limiters. It is evident that the Min-mod limiter with $\beta = 1$ predicts the highest spatial error, while $\beta = 4$ possesses the lowest spatial error.

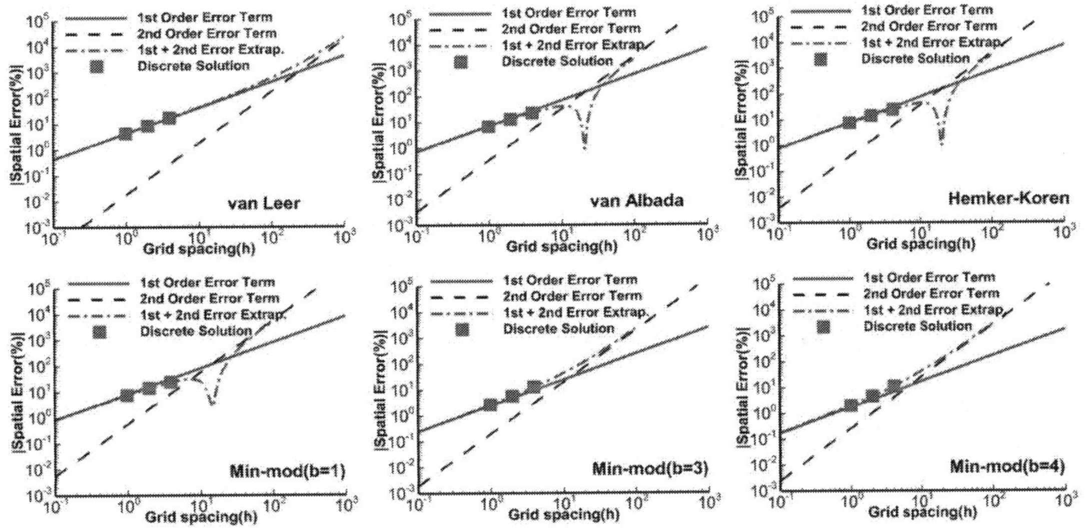

Figure 3: Grid-convergence error in drag coefficients with different limiters as grid refined.

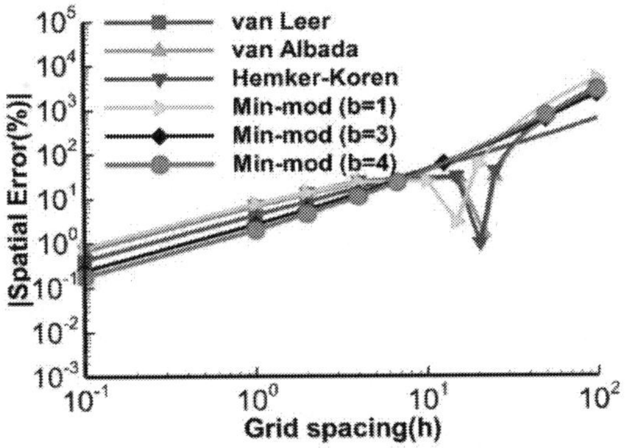

Figure 4: Comparison of mixed-order spatial errors of different limiters for the incident shock on flat plate.

The separation bubbles due to the interactions of the shock wave with the boundary layer predicted by different limiters are shown in Fig.5, where the streamlines around the separation bubble are also plotted. The changes of the boundary layer along the entire flat plate can also be seen from Fig.5, especially in the region of separation bubble. As the dissipative characteristic of limiter gets stronger, the predicted size of separation bubble becomes smaller. The most dissipative Min-mod limiter with $\beta = 1$ predicts the smallest separate region. The sizes of separation region predicted using the Min-mod limiters of $\beta = 3, 4$ are larger than those predicted by implementing van Leer limiter, van Albada limiter and Hemker-Koren limiter.

From the above discussions, it can be concluded that, for discontinuity-containing flows, the spatial error, i.e. the solution accuracy, is mainly determined by the dissipation level of the limiters adopted: the less dissipation for the limiter, the more accurate the solution.

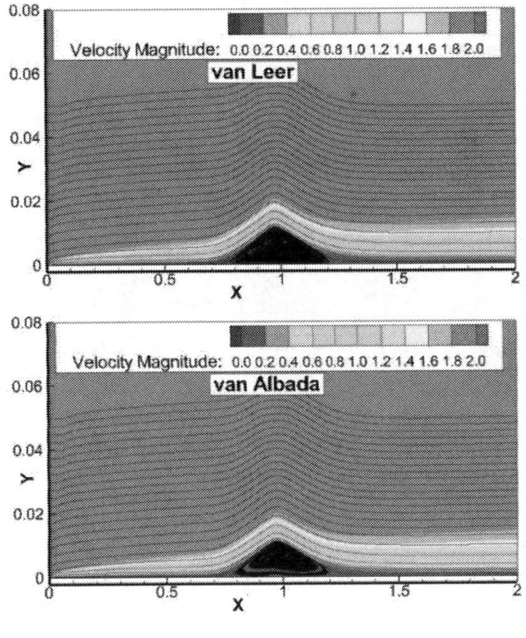

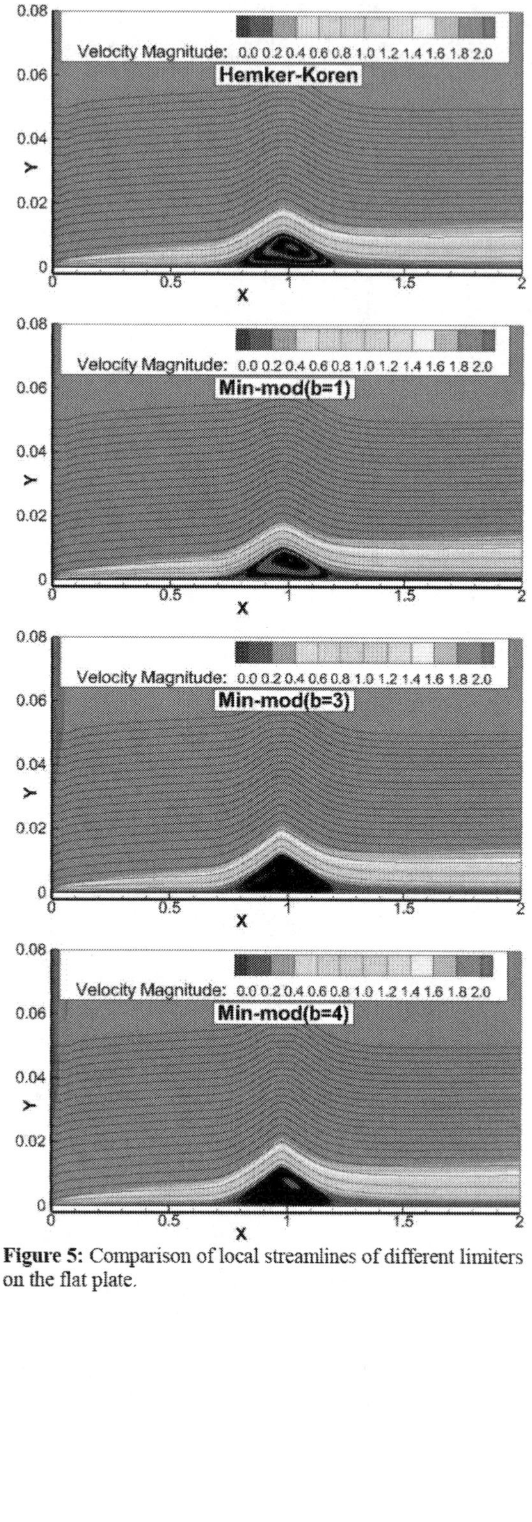

Figure 5: Comparison of local streamlines of different limiters on the flat plate.

5 Conclusion

The properties of several frequently used MUSCL-type limiters, including van Leer limiter, van Albada limiter, and Hemker-Koren limiter functions and Min-mod limiter are assessed together. Grid independent studies based on the mixed-order spatial error estimator are conducted to investigate the influence of limiter's dissipation/compression level on the solution accuracy for the first time in available public publications. Based on the mixed-order error estimating method, the grid-independent solutions are of first-order accuracy in shockwave-containing flows and of second-order accuracy on the coarser grid. The spatial errors are of different levels with different limiters which possess different intrinsic dissipative/compressive properties. These intrinsic dissipative/compressive features play an important role in the solution accuracy in predicting strong shockwave-containing flows.

References

1. S. Pirozzoli, F. Grasso, T.B.Gatski, Phys. Fluids **16,** 530–545 (2004).
2. H. Zhao, P. Balakumar, AIAA J. **43,** 1034–1041 (2005).
3. H.C. Yee, N.D. Sandham, M.J. Djomehri, J. Comput. Phys. **150,** 199–238 (1999).
4. B. Van Leer, Comput. Phys. **136,** 101–136 (1979).
5. C.J. Roy, AIAA J. **41,** 595–604 (2003).
6. C.J. Roy, W.L. Oberkampf, M.A.McWherter-Payne, AIAA J. **41,** 1944–1954 (2003).
7. C.J. Roy, M.A. McWherter-Payne, W.L. Oberkampf, AIAA Paper. **2000–2550,** (2000).
8. G. Li, Z. Zhou, AIAA Paper. **2010–4549,** (2010).
9. K.H. Kim, C. Kim, O.H. Rho, J. Comput. Phys. **174,** 38–80 (2001).
10. M-S. Liou, C.J. Steffen, J. Comput. Phys. **107,** 23–39 (1993).
11. W.K. Anderson, J.L. Thomas, B. Van Leer, AIAA J. **24,** 1453–1460 (1986).
12. S. Spekreijse, Math. Comput. **49,** 135–155 (1987).
13. P.J. Roache, J. Fluids Eng. **116,** 405–413 (1994).
14. P.J. Roache, Computing in Science Engineering 8–9 (1998).
15. R.J. Hakkinen, I. Greber, L. Trilling, S.S. Abarbanel, NASA Memo **2–18–59W,** (1959).

https://doi.org/10.1051/matecconf/201925702002

Identification of Natural Bamboo Fiber and Regenerated Bamboo Fiber by the Method of Modified near Infrared Spectroscopy

Weidong Li [1], Zhixiong Li [1] and Xinhou Wang [2]

[1] Shanghai Institute of Quality Inspection and Technical Research, Shanghai 200040, P. R. China
[2] Donghua University, Shanghai 201620, P. R. China

Abstract. In this work, the applicability and effectiveness of identification of natural bamboo fiber and regenerated bamboo fiber by near infrared spectroscopy are investigated. A discrimination model based on Ward's algorithm and Hierarchical Cluster Analysis is pretreated by the first derivative and vector normalization, which can be used to distinguish natural bamboo fiber and regenerated bamboo fiber. In addition, the near infrared spectra model exhibits a high accuracy with an effective identification of the double fibers.

1 Introduction

Bamboo fiber, as a kind of renewable, biodegradable, low-priced and abundant cellulosic fibers, it has been widely used in textiles and clothing by virtue of their excellent performance of physics and chemistry [1-4]. Bamboo fiber consists of natural bamboo fiber and regenerated bamboo fiber [5]. Natural bamboo fiber is fabricated by removing plant tissues without cellulose based on the mechanical and physical methods [6]. The fabrication of regenerated bamboo fiber is similar to the common viscose fiber, which is prepared by chemical spinning solution [1].

The structure and function is different between natural bamboo fiber and regenerated bamboo fiber due to their different manufacturing process. At present, some methods and papers on the identification of bamboo fibers have been put forward, but their applicability and effectiveness are unsatisfactory [7-9]. In this paper, natural bamboo fibers and regenerated bamboo fiber produced by the main domestic manufacturers are collected widely, and the accurate identification of these two kinds of fibers is realized by the near infrared spectroscopy (NIR) technology in virtue of the cluster analysis model.

2 Experimental section

2.1 Materials

The materials used in the experiments are listed in Table 1. Natural bamboo fibers supplied by Sichuan Banbo Bamboo Development Co., Ltd and Fujian Zhenghe Bambootextile Co., Ltd; bamboo pulp viscose fibers supplied by Jigao Chemical Fiber Co., Ltd of Hebei provinceand Suzhou Shenboo Textile Co., Ltd.

Table 1. Materials used for the NIR spectra analysis

Fiber samples	Number of the modeling sample set	Number of the correction sample set
Natural Bamboo fibers	15	4
Regenerated bamboo fiber	10	4

(Note: fibers used for the NIR analysis are numbered in random order)

2.2 Instruments

MPA multi-functional Fourier transform near infrared spectrometer (Bruker Co.). Diffuse reflectance integrating sphere attachment (built-in gold-plated background), gold-plated integrating sphere diameter is 10 cm, measurement window diameter is 2 cm, PbS detector. Scanning resolution is 4 cm^{-1}, the number of scans is 32, spectral range 12 000-4000 cm^{-1}, the OPUS software.

3 Results and discussion

The NIR spectra of the natural bamboo fiber and regenerated bamboo fiber measured by diffuse reflection method are shown in Figure 1A and 1B, respectively.

https://doi.org/10.1051/matecconf/201925702002

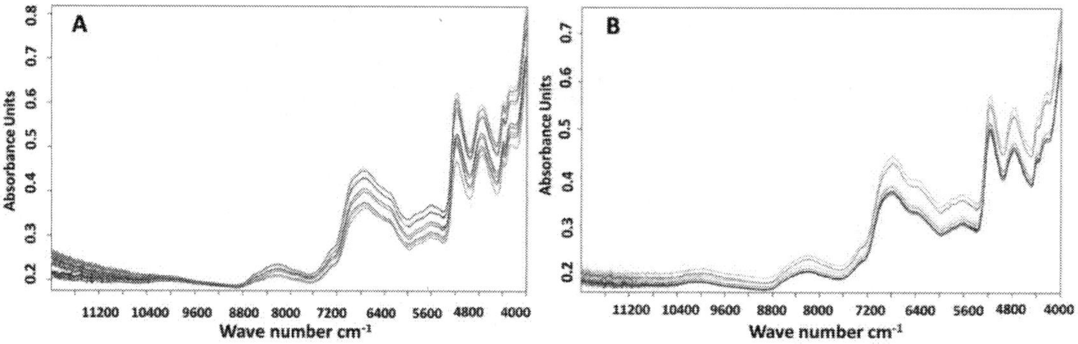

Figure 1. NIR spectra of (A) natural bamboo fiber and (B) regenerated bamboo fiber.

The NIR spectra of natural bamboo fiber and regenerated bamboo fiber are similar, the obvious absorption peaks at 4000～8800 cm^{-1} can be observed. Therefore, the first derivative is proposed to eliminate the influence of baseline and background with 10 smooth points, which can intensify the character of respective NIR spectra and keep the effective information. Vector normalization is adopted to eliminate the interference caused by the locations, thickness, manufacturers and producing areas guaranteeing a reliable NIR spectra information.

Then, Ward's algorithm model is employed to calculate the distance based on a view of variance analysis and heterogeneous definition rule. As an excellent methodology to distance's measure, Ward's algorithm model displays an optimized result with a smaller congeneric variance and a larger heterogeneous variance.

Finally, a discrimination model based on Ward's algorithm and Hierarchical Cluster Analysis is built with the help of optimized NIR spectra [10]. As a result, the identification of natural bamboo fiber and regenerated bamboo fiber is realized (Figure 2).

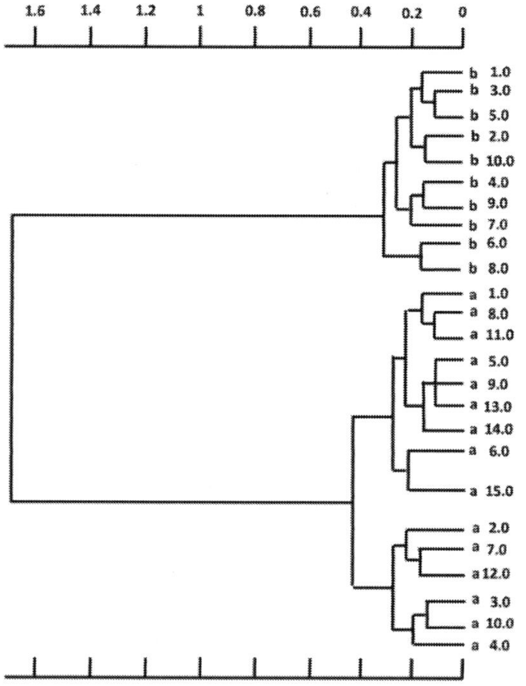

Figure 2. Cluster analysis model of natural bamboo fibers and regenerated bamboo fiber. (Note: a represents natural bamboo fiber, b represents regenerated bamboo fiber. The coordinate value represents the value of the distance between groups)

An accuracy and effective verification of the discrimination model is necessary. The NIR spectra of fiber sample which is pretreated by the first derivative and vector normalization is shown in Figure 3. The verified analysis model and overall verified results are displayed in Figure 3 and Table 2, respectively.

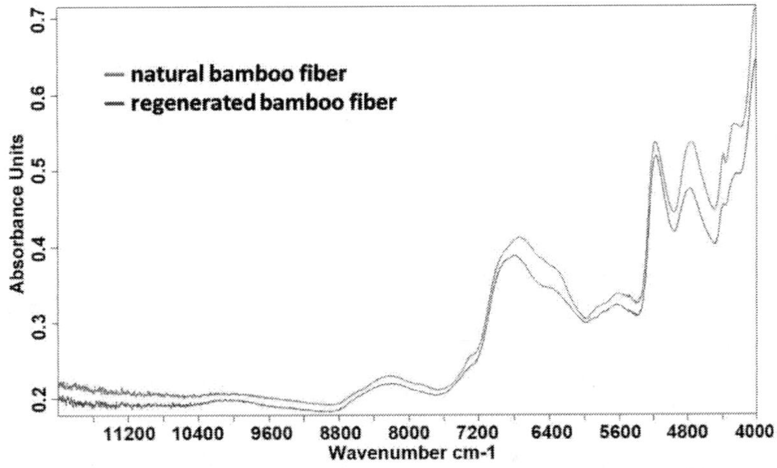

Figure 3. Correctional NIR spectra of the fiber sample set (part for example).

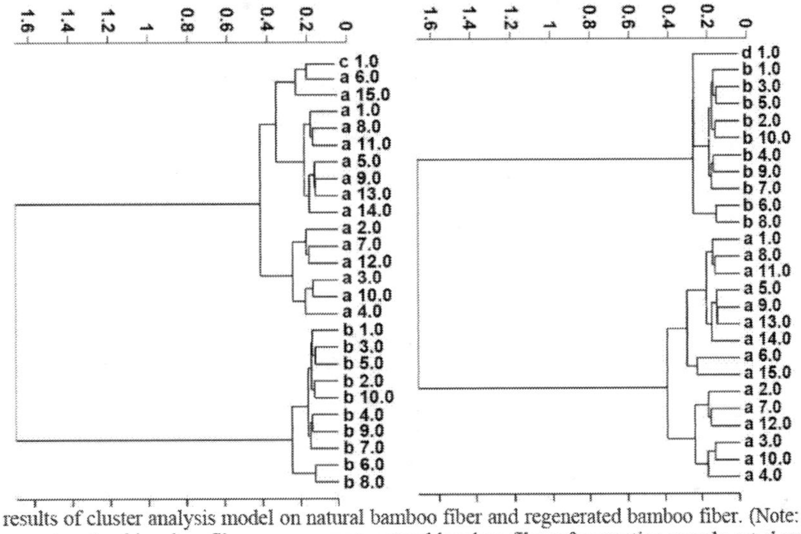

Figure 4. Verified results of cluster analysis model on natural bamboo fiber and regenerated bamboo fiber. (Note: a represents natural bamboo fiber, b represents natural bamboo fibers, c represents natural bamboo fiber of correction sample set, d represents regenerated bamboo fiber of correction sample set. The coordinate value represents the value of the distance between groups)

Table 2. Overall verified results due to NIR spectra and cluster analysis model

Samples	Number of the correction sample set	Right results	Wrong results	Model accuracy (%)
natural bamboo fibers	4	4	0	100
regenerated bamboo fiber	4	4	0	

The identification of the two fibers can be realized with relatively high accuracy based on the results in figure 4 and table 2. In addition, the cluster analysis model established by Hierarchical Cluster Analysis can identify natural bamboo fiber and regenerated bamboo fiber simply and effectively without any damage to the fiber samples. And in this system, the accuracy increases with the growing of representativeness and class of fiber samples, which shows a great potential in the field of sensors and detection.

4 Conclusion

In a word, a discrimination model based on Ward's algorithm and Hierarchical Cluster Analysis is pretreated by the first derivative and vector normalization, which achieve the effective identification of natural bamboo fiber and regenerated bamboo fiber with an excellent operability, veracity and detection speed. The method of modified near infrared spectroscopy can be widely used in the field of the identification between similar natural and regenerated fibers, showing an extensive application for the sensor, identification and detection.

References

1. S. Y. Sui and R. Q. Li, A study of structure and performance of bamboo fibers, *Journal of Textile Research*, 24 (2016), 535-536.
2. S. Y. Lee, S. J. Chun, G. Doh, et al., Influence of chemical modification and filler loading on fundamental properties of bamboo fibers reinforced polypropylene composites, *Journal of Composite Materials*, 43 (2009), 1639-1657.
3. S. B. Ma and P. Y. Wu, Identification and performance analysis of bamboo pulp fiber and viscose fiber, *Wool Textile Journal*, 38 (2010), 42-46.
4. T. Zhang, W. B. Bao and J. Y. Yu, Study on identification of bamboo fiber, *Journal of Textile Research*, 25 (2004), 28-29.
5. W. D. Li, Research progress in determination methods of bamboo fiber, *China Fiber Inspection*, 359 (2010), 61-64.
6. L. D. Cheng, X. L. Xu and J. H. Lao, An analysis on the morphological structure and property of bamboo fiber, *China Textile Leader*, 5 (2003), 101-103.
7. Q. B. Zhou and J. L. Chen, The performance difference between bamboo pulp fibre and bamboo fiber, *Advanced Textile Technology*, 2 (2007), 1-3.
8. J. P. Zhou, Study on the structure and property of original bamboo fiber and bamboo pulp fiber, *Shanghai Textile Science & Technology*, 34 (2006), 59-60.
9. J. J. Sun, J. Y. Tian and Z. Y. Gu, Comparison of structure and thermal property between bamboo fibers and regenerated bamboo fibers, *Journal of Tianjin Polytechnic University*, 25 (2006), 37-40.
10. J. X. Liu, Practical technology of the near infrared spectra analysis, Beijing: Science Press, 2007.

MATEC Web of Conferences **257**, 02003 (2019) https://doi.org/10.1051/matecconf/201925702003
PMAE 2018

Thermal Characteristics Analysis for Motorized Spindle System of CNC Machining Center

Xiaolei Deng[1,2,3], Xinghui Zhang[1], Mucheng Zhang[1], Yibo Zhou[1], Huan Lin[1] and Jianchen Wang[1,3]

[1]*Key Laboratory of Air-driven Equipment Technology of Zhejiang Province, Quzhou University, Quzhou 324000, China;*
[2]*Key Laboratory of 3D Printing Process and Equipment of Zhejiang Province, Zhejiang University, Hangzhou 310027, China;*
[3]*Zhejiang Yonglida CNC Technology Co., Ltd., Quzhou 324000, China*

Abstract. Based on the comprehensive analysis of the heat sources of the motorized spindle system, the thermal loads, including the heat generation of bearing friction and the electromagnetic loss of the built-in motor, are carried out for a machining center motorized spindle system. And then, the convective heat transfer coefficients of the whole spindle system are analyzed. The thermal characteristics of the motorized spindle system are calculated by finite element analysis. The steady state temperature field distribution of the motorized spindle is obtained. It provides some references for improving the thermal characteristics of the motorized spindle and reducing the difficulty of thermal error compensation.

1 Introduction

The motorized spindle system will produce deformation due to stress in the process, in addition to large amounts of heat, which can generate thermal deformation. These thermal deformations can not only short the working life of the spindle, but also affect the processing accuracy of the machine tool. A large number of studies have shown that in the factors that affect the accuracy of machine tool, the thermal error caused by external environment and internal heat sources are the largest error source of precision machining machine such as CNC(Computer Numerical Control) machine tools, which accounts for 40-70% of the total manufacturing error [1, 2].

Therefore, it is necessary to analyze the temperature field of the motorized spindle system and grasp the thermal characteristics of the motorized spindle system, so as to find a way to effectively reduce the temperature rise of the spindle, and reduce the thermal errors of the spindle system.

In this paper, a machining center's motorized spindle system was taken as an example to study its heat and cooling mechanism. Using finite element software ANSYS, the steady temperature field of the motorized spindle system were obtained.

2 Calculation of the thermal load of the motorized spindle system

There are two main sources of heat source in the motorized spindle system, including the heat due to loss of the built-in motor and the heat due to friction of the spindle-bearing.

2.1 Heat calculation of built-in motor

The heat generated by the motor in the motorized spindle system can be divided into heat generation of rotor and heat generation of the stator. The power of the motorized spindle system from the input power P_1 can be calculated by Eq.1

$$P_1 = 3U_1 I_1 cos\varphi_1 \tag{1}$$

where, U_1 is phase voltage of stator winding, I_1 is phase current of stator winding, $cos\varphi_1$ is power factor of stator winding.

After motorized spindle system take power P_1 suction, one part is consumed on the stator winding resistance (r_1), i.e. copper loss P_{Cu1} which canbe calculated from

$$P_{Cu1} = 3I_1^2 r_1 \tag{2}$$

Another part is consumed on the stator iron core, i.e. iron loss P_{Fe} which canbe calculated from

$$P_{Fe} = 3I_0^2 r_m \tag{3}$$

So the total power loss of the stator (P_d) is P_{Cu1} and P_{Fe}.

The heat generation rate of the stator [3]

$$\overline{q_d} = \frac{P_d}{V_d} \tag{4}$$

where, V_d is the volume of the stator.

In addition, the rest of the power is transferred from the stator to the rotor through the electromagnetic induction, which is called the electromagnetic power P_M can be calculated by Eq.5

$$P_M = P_1 - P_{Cu1} - P_{Fe} \tag{5}$$

Due to the low frequency of the rotor in normal operation, the iron consumption of the rotor is very small, usually negligible, so the rotor power loss P_z can be equivalent to the copper consumption of the rotor.Thus the heat generation rate of the rotor is

$$\overline{q_z} = \frac{P_z}{V_z} \tag{6}$$

where, V_z is the volume of the rotor.

2.2 Calculation of heat for bearing

According to Eq.7 [1], the friction moment of the bearing is mainly composed of two parts: the viscous friction moment and the load friction moment, and the calculation formula as following

$$M = 10^{-7} f_0 (vn)^{\frac{2}{3}} d_m{}^3 + f_1 P_1 d_m \tag{7}$$

where, M is the friction torque of bearings, f_0 is depends on the type of bearing, v is motion viscosity of lubricants at working temperature, n is rotating speed of inner ring of bearing, d_m is circle diameter of bearing, f_1 is coefficient related to bearing type and load, P_1 is calculation load determining the friction torque of bearing.

In addition, the spin motion of the rolling body is an important factor in heating. And the frictional moment in the contact area between the rolling body and the inner ring is [4]

$$M_{Si} = \frac{3\mu_{Si} Q_i a_i \varepsilon_i}{8} \tag{8}$$

where, μ_{Si} is the friction coefficient of the rolling element and the inner ring raceway, a_i is Hertz contact ellipse long axis of rolling element and the inner ring raceway, Q_i is the normal contact load of the rolling element and the inner ring raceway, ε_i is the second complete integral of Hertzian contact ellipse for rolling element and the inner ring raceway.

The friction power is generated by the bearing as following [5]

$$H = \frac{2\pi n_1}{60} M + \omega_{Si} M_{Si} \tag{9}$$

where, n_1 is the rotation speed of the spindle, ω_{Si} is spin angular velocity at the contact part between the roller and the inner raceway.

The heat generation rate of the bearing is defined as Eq.10 [6]

$$\begin{cases} q_f = \dfrac{H_f}{V_f} \\ q_a = \dfrac{H_a}{V_a} \end{cases} \tag{10}$$

where, q_f is the heat generation rate of the front bearing, q_a is the heat generation rate of the rear bearing, H_f is the friction power of the front bearing, H_a is the friction power of the rear bearings, V_f is the volume of the front bearing, V_a is the volume of the rear bearings.

3 Calculation of the thermal boundary condition of the motorized spindle system

3.1 Convective heat transfer coefficient between stator cooling water jacket and cooling water

The stator outer ring of the motorized spindle system has a rectangular spiral cooling water jacket to cool the stator and the outer shell. The Nusselt criterion equation, under forced convection, convective heat transfer coefficient can be alculated by Eq.11

$$a = \frac{\mathrm{Nu} \cdot \lambda}{L} \tag{11}$$

where, λ is the thermal conductivity of the ambient fluid, Nu is Nusselt number, L is the equivalent length of spiral cooling water jacket.

The Nusselt number of cooling water in laminar state can be calculated by Eq.12

$$\begin{cases} \mathrm{Nu} = 1.86\left(\mathrm{Re\,Pr}\dfrac{D}{L}\right)^{\frac{1}{3}} \left(\dfrac{\eta_f}{\eta_w}\right)^{0.14} \\ \mathrm{R_e} = \dfrac{v_1 \cdot D}{v_f} \end{cases} \tag{12}$$

where, Pr is the Prandtl number, η_f is fluid dynamic viscosity, η_w is dynamic viscosity of the fluid under the cooling water jacket wall temperature, v_1 is the flow rate of cooling water, D is stereotyped size of geometric features, v_f is kinematic viscosity of fluid in qualitative temperature.

The calculation method of convective heat transfer coefficient between the annular groove and the cooling water of the front bearing block is the same as the method above.

3.2 Convective heat transfer coefficient between the cooling gas and the stator

The average speed of cooling gas between the rotor and the stator is defined as follows

$$\bar{u} = \sqrt{\bar{u}_{ax}^{\,2} + \bar{u}_{rad}^{\,2}} \tag{13}$$

The average velocity of the axial cooling gas through the air gap between the stator and rotor is defined as follows

$$\bar{u}_{ax} = \frac{v_g}{A_{ax}} = \frac{v_g}{\frac{\pi}{4}\left(d^2{}_{stator} - d^2{}_{rotor}\right)} \tag{14}$$

where, v_g is the volume flow rate of cooling gas, A_{ax} is the axial cross-sectional area of gap, d_{stator} is the stator inner circle diameter, d_{rotor} is the rotor outer circle diameter.

In addition, the average velocity of the cooling gas is subjected to the shear action of the circumferential direction of the stator and the rotor, resulting in the tangential mean velocity.

$$\bar{u}_{rad} = \frac{v_{sur}}{2} = \frac{\omega_{rotor} d_{rotor}}{4} \tag{15}$$

where, v_{sur} is the velocity of the rotor surface.

3.3 Convective heat transfer coefficient of the rotor end

When the spindle is running, the rotor end is mainly convection and radiation heat transfer with the surrounding cooling air. The compound heat transfer coefficient is as follows [7]

$$a_t = 28\left(1+\left(\frac{3\pi n_1 d_a}{400}\right)^{0.5}\right) \tag{16}$$

where, a_t is the convection heat transfer coefficient at the rotor end, d_a is the average diameter of the rotor end.

3.4 Convective heat transfer coefficient between bearing and lubricating gas

The heat transfer coefficient between the bearing and the lubricating gas can be determined by Eq.17

$$\alpha = \left(c_0 + c_1 u^{c_2}\right) \tag{17}$$

where, c_0, c_1 and c_2 are determined by the constant of the test; u is the average velocity of the compressed gas in the bearing.

The average velocity of compressed air in a rotating bearing can be obtained by the axial and tangential air flow.

$$u = \left[\left(\frac{V_1}{A_a}\right)^2 + \left(\frac{\omega d_m}{2}\right)^2\right]^{\frac{1}{2}} \tag{18}$$

where, V_1 is the flow of the actual gas of the bearing; ω is the rotational angular velocity of the spindle, A_a is the area when the axial air flow passes through the bearing, d_m is the average diameter of the inner and outer ring of the bearing.

3.5 Heat transfer coefficient between the outer shell and the surrounding air

When the motorized spindle system is running, the natural convection heat transfer with the surrounding air is caused by the surface temperature of the shell is higher than the ambient air. Meanwhile, the spindle shell also radiate heat with the surrounding air. The heat transfer coefficient of the composite heat transfer is α_s

$$\alpha_s = \alpha_c + \alpha_r \tag{19}$$

where, α_c is convective heat transfer coefficient which is converted from radiation, α_r is natural convection heat transfer coefficient, α_s is compound heat transfer coefficient.

4 Thermal analysis of motorized spindle system

Taking one motorized spindle system as the research object, the thermal characteristics of the spindle system are analyzed according to the load and boundary conditions calculated by the above thermal characteristic models. It is known that the temperature of the external environment is 22 °C, the spindle speed is 42000 rpm, the cooling water flow is 1.07 L/min, and the gas supply pressure of the oil lubrication system is 0.23 MPa. And Ansys workbench software is employed to gain the steady temperature field distribution of the spindle system, as shown in Figure 1.As can be seen from figure 1, the maximum temperature appears at the front bearing is about 154.06 °C, and the maximum temperature of shaft is about 140.38 °C.

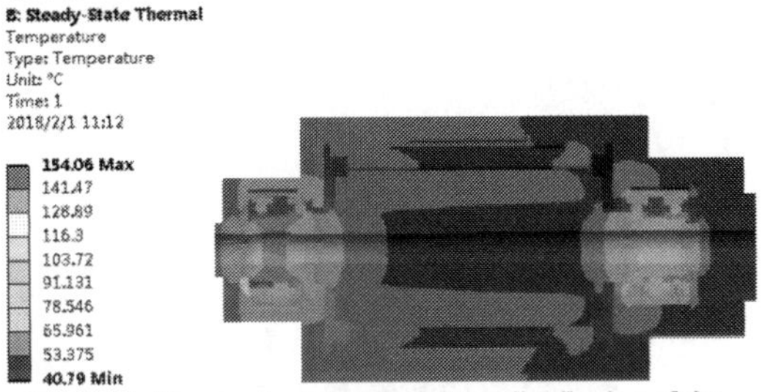

Figure 1. The steady state temperature distribution of the motorized spindle system

5 Conclusions

In this paper, a thermal load and boundary condition calculation model of motored spindle system for one machining center was established, and Ansysworkbench was employed to simulate and solve. Through the analysis, the steady state temperature distribution of the motorized spindle system was gotten.The research in this paper will provide some valuable reference for the thermal structure design and thermal error compensation of the motorized spindle system.

Acknowledgment

This research was financially supported by National Natural Science Foundation of China (No. 51605253), Zhejiang Provincial Natural Science Foundation of China (No. LY16E050011) and Quzhou Science and Technology Planning Project（2016Y003&2015Y007）.

References

1. DENG X L. Analysis Techniques of Thermal Characteristics for CNC Machine Tool Spindle System [M].Hangzhou: Zhejiang University Press, 2017. (in Chinese)

2. Mayr J, Jędrzejewski J, Uhlmann E, et al. Thermal Issues in Machine Tools [J]. CIRP Annals-Manufacturing Technology, 2012, 61(2): 771-791.
3. Huang X M.FEA of Thermal Properties for High Speed Motorized Spindle [D].Guangzhou: Guangdong University of Technology, 2003. (In Chinese)
4. Deng S E, Jia Q Y. Design Principle of Rolling Bearing [M].Beijing: China Standards Press, 2008.(In Chinese)
5. Brauer J R, Mayergoyz I D. Finite-element Computation of Nonlinear Magnetic Diffusion and Its Effects When Coupled to Electrical, Mechanical, and Hydraulic Systems [J]. IEEETransactions on Magnetics, 2004, 40(2): 537-540.
6. Guo J. Research on Motorized Spindle's Thermal Properties Based on Thermal Contact [D].Guangzhou: Guangdong University of Technology, 2005. (In Chinese)
7. Ding S N. Heating and Cooling of Large Motor [M].Beijing: Science Press, 1992.

The Optimization of FDM 3D Printer`s Structure Based on Finite Element Analysis

Zhenhai Huang[1], Tingchun Shi[1], Xiuyan Yue[2]

[1]*Hangzhou Dianzi University, Hangzhou, China*
[2]*Hangzhou Dianzi University Library, Hangzhou, China*

Abstract. In view of the forming accuracy of the 3D printer, the paper, from the structure of the printer itself, discussed the influence on the precision of the printer via improving the structure of 3D printer. A typical 3D printer, such as a common type of FDM, was selected, and SOLIDWORKS software was used for solid modeling, after establishing the model, the finite element analysis was carried out on the whole structure model to obtain the deformation characteristics and the main vibration model of the printer. On the basis of finite element analysis, the influence of the level shaft bearing diameter, material properties and symmetry of the parts on the overall mechanical properties of the printer was discussed respectively. On this basis, the overall structure of the machine has been optimized and achieved remarkable results.

1 Introduction

Fused deposition modeling (FDM) technology 3D printer has been widely applied in the market due to its high forming precision, low price and easy to obtain printing materials. Although FDM technology tends to mature, there are still many limitations in the process of forming and manufacturing such as molding precision, material type and so on. The factors that affect the forming precision of FDM type 3D printer can be divided into layered software, control algorithm, mechanical structure and so on. At present, the printer mechanical structure research mainly focus on the nozzle, column and other single parts of the analysis, relatively, on the overall structure of the 3D printer, is lack of integrity, and the research depth is not enough[1, 2].

In this paper, a common PrusaI3 series 3D printer is selected as the research object. Firstly, the overall model of the 3D printer is modeled. Secondly, the static mechanical analysis and dynamic mechanical analysis of the established printer model were carried out by using ANSYS Workbench software to get the deformed cloud chart, natural frequency and modal shape of the printer model. Finally, the result of the finite element analysis is used as a reference to optimize the mechanical structure of the printer.

2 3D printer finite element model establishment

2.1 The establishment of CAD model for 3D printer

The PrusaI3 series is one of the more popular FDM 3D printers available on the market[3, 4]. The advantages of these printers are their simple mechanical structure and high precision. The printer's nozzle system includes the feeding and melting function, during the operation of the printer, which moves along the fixed shaft bearing. The modeling platform is located on the bottom shaft bearing of the 3D printer, and the motion direction of the platform is perpendicular to the motion plane of the nozzle system.

In order to accurately reflect the real situation of the printer, the SOLIDWORKS software is used to model the overall structure of the 3D printer in accordance with its actual size. In the process of modeling, according to the degree of impact on the subsequent finite element analysis, the printer structure is simplified. The two side screw bearing of 3D printer can be replaced by the same standard cylindrical shaft bearing, the synchronous belt is omitted and the stepper motor is simplified with a motor model of the same weight and shape. The simplified mode of the 3D printer overall structure and its main components are illustrated as shown in Figure 1.

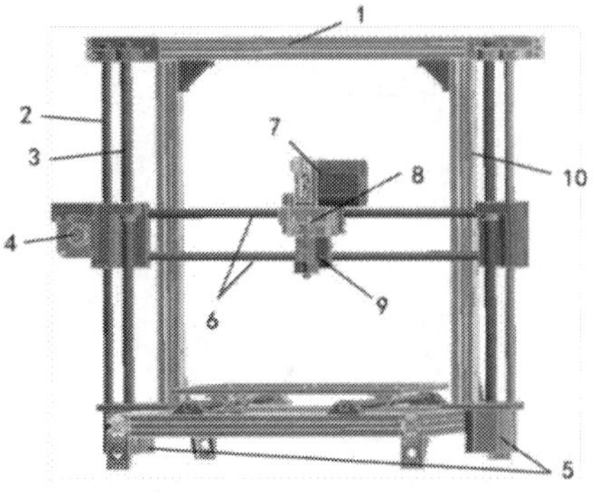

Figure 1. Simplified model of 3D printer

1.beam, 2. vertical shaft bearing, 3. simplified screw, 4, 5. stepper motor, 6. level shaft bearing, 7. feed motor, 8. measuring points, 9. nozzle melting part, 10. columns.

2.2 The material attribute setting of the model

The rack part of 3D printer is mainly made up of aluminum alloy material. The type of aluminum alloy is 6060-T5[5],the link between the two aluminum profiles is a fixed tripod, and the material is also aluminum alloy. In order to facilitate calculation, we set up the material property uniformly. The diameter of vertical and level shaft bearing of the 3D printer is 8mm, and the material is GCR15 bearing steel[6]. In the printer, the connecting part of the motor, the fixed nozzle part, and the part of the fixed level shaft bearing are the components with the material PLA[7]. The properties of GCR15 alloy steel materials, MJ-6-2020 aluminum alloy materials, PLA and other materials, as shown in Table 1.

Table 1. 3D printer frame main material properties

Type of material	Density(Kg/m^3)	Poisson ratio	Modulus of elasticity
bearing steel	7810	0.3	208Gpa
Aluminium alloy	2700	0.33	70Gpa
PLA	250	0.402	3500Mpa

2.3 Simplification of the contact surface of the model

The processing method of the mechanical contact surface is important for the mechanical structure; the contact surface of the printer model can be roughly divided into fixed contact and non-separation contact. Fixed contact is suitable for objects with the same way of motion, the two objects cannot be separated or slipped. For the non-separation contact, it could have gaps that can rotate, but not allow lateral movement.

The aluminum section is linked by a tripod and is screwed through a M5*8 type screw. The size of the pre-tightening force is 6.4kN. The frame of the 3D printer structure can be well fixed, and the contact with fixed screws on the printer structure can be treated as fixed contacts. For the printer nozzle system and the forming platform, the contact surface connected with the level shaft bearing is in a slippery state during the movement of the printer; however, in the static situation, there is a large sliding resistance on both sides and the resistance of the motor itself and the movement is difficult. In the subsequent finite element analysis, it is more convenient to treat it as a non - separation contact. In addition, in the analysis of the whole 3D printer structure, in order to simplify the calculation, the stepper motor model under the forming platform can be removed.

3 Finite element analysis of 3d printer structure

Before the computation of finite element analysis, the mesh of the printer model needs to be divided, which is an important part of finite elements[8, 9]. The quality of mesh will affect the transmission of force in the process of finite element calculation, which may ultimately affect the accuracy of calculation results. Under the default conditions of ANSYS workbench software, the quality of the grid obtained is relatively rough, and the grids between the various components have a larger difference. There are more grid division methods in the finite element analysis software. The mechanical part has a regular shape. When the mesh is optimized, the hexahedron-dominant method is used to optimize the mesh. The average quality of the mesh obtained after optimization is 73.6%. The main purpose of finite element analysis is to explore the stability of 3D printer's mechanical structure, and the finite element analysis of the whole 3D printer structure can be mainly divided into the statics analysis and modal analysis.

3.1 Statics analysis of 3D printer structure

In the forming process of the 3D printer, the direct forming error comes from the stability of the nozzle system[10]. The nozzle system is divided into the melt forming system and the feeding system. The melt forming system is responsible for the melting and wiring of the molding material. Feeding system is located above the melt forming system and consists of feeding stepper motor and feeding model, which is responsible for material feeding work. The supporting part of the whole nozzle system is the two level shaft bearings which is connected to the nozzle system by a linear bearing.

The main purpose of static analysis is to analyze the reliability of the mechanical structure of the 3D printer and to discuss the potential defects in the structure of the printer. In the statics analysis of the finite element, the bottom of the four aluminum alloy tripod under the printer is fixed. After completing the boundary condition setting and grid optimization of the printer, the static analysis results of the printer under its own gravity are shown in Figure 2.

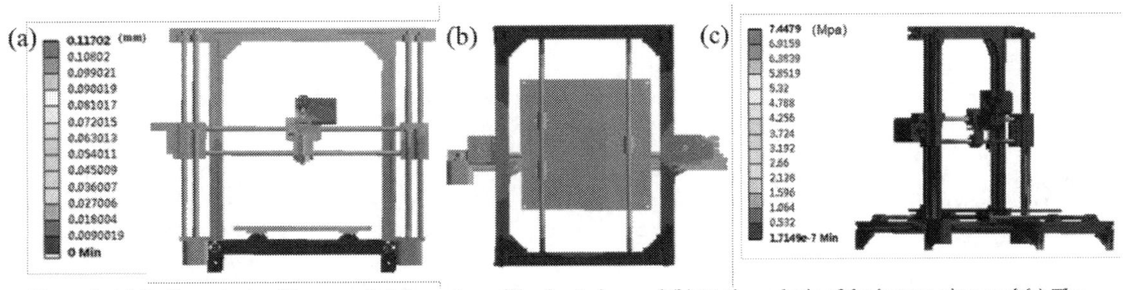

Figure 2. (a) Static analysis of the overall deformation of the front view and (b) Static analysis of the bottom view, and (c) The pressure distribution of static analysis

In the static analysis results of finite element, Fig. 2(c) is the pressure distribution chart, from which it can be seen that the main load-bearing parts of the printer are distributed in the contact parts between the bottom bracket and the column, as well as the contact part of the modeling platform and the nozzle system with its optical axis. The overall deformation of the 3D printer as shown in Fig.2(a)， the printer structure has the largest amount of deformation at the feeding system, the tendency of deformation is to extend outwards from the fixation of the level shaft bearings, and the shape variable presents an increasing form, and the maximum deformation reaches 0.117mm. As can be seen from Fig.2(b), the deformation of the forming platform shows an asymmetrical trend. The deformation of the lower side is about 0.01mm, while the upper side reaches 0.06mm. At present, the nozzle diameter of the FDM type 3D printer on the market is generally 0.3-0.4 mm, and the maximum precision of the printer can reach 0.1 mm. In addition, in actual conditions, since the printer's molten nozzle is not firmly fixed, there may be a small swing in the forming of the printer. Considering the influence of these factors, the overall structural deformation of the 3D printer in static analysis will have a great influence on the molding error.

3.2 Modal analysis of 3D printer structure

In the working environment of the 3D printer, the excitation is dynamically loaded, and it should be avoid coupling the natural frequency with the external excitation[11-13]. When the mechanical structure of the coupling frequency is reduced, it can not only improve the stability of the printer itself, but also reduce the load of the machine itself and extend its service life. In the work of 3D printer, the main source of excitation is generated by stepper motor, which is at a relatively low level. In the modal analysis of the 3D printer, the vibration characteristics and natural frequency of the first to fourth orders are selected, and the modal analysis results of the whole structure of the printer are shown in Fig.3

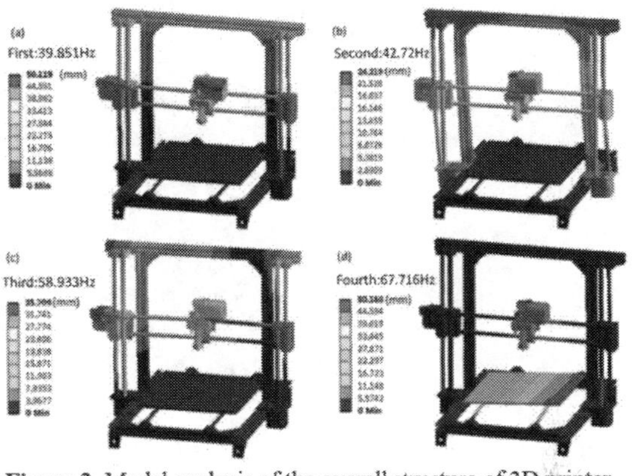

Figure 3. Modal analysis of the overall structure of 3D printer

It can be seen from the results of the modal analysis that the range of modal frequencies from the first to fourth orders is within 39-68hz. When the frequency is low, the printer has obvious vibration phenomenon, such as the both sides of printer columns, beam parts, and nozzle system. When these analysis and optimization are carried out, we should make reasonable optimization for these parts.

4 Optimal design of mechanical structure

In general, the principle of mechanical structure optimization is to improve the natural frequency of the mechanical structure and avoid the occurrence of frequency coupling[14, 15]. According to the analysis of the above finite element, in the discussion of the optimization of the printer's structure, the structure of the 3D printer can be divided into the entire nozzle system and the remaining part of the rack. The maximum deformation of the statics analysis and the natural frequency of the modal analysis are considered as the criteria for the optimization analysis.

5 Optimization analysis of nozzle system

5.1 The influence of structure improvement on the stability of nozzle system

The nozzle system of the prusaI3 type 3D printer can be divided into feeding system and heating forming system. According to the characteristics of modal analysis, in order to improve the natural frequency of the nozzle system, in the case of the same stiffness, it should be appropriate to reduce the overall quality of the nozzle system, however, the quality of the nozzle system is largely due to the stepper motor in the feeding system, whose function is to provide feed power for the feeding system. In the process of improvement, the nozzle system of feeding system and heating molding system are divided, and it is to use the principle of remote feeding. The feeding system is connected to the heating system by the engineering plastic conduit, when the printer works, the printed material is sent through the tube to the heating melting system. The feeding system can be placed as a separate part next to the printer or a part fixed to the frame of the 3D printer. In order to verify the feasibility of the method, the nozzle is firstly separated, and then the nozzle system of the printer before and after separation is calculated to do the statics analysis. The resulting static analysis of the deformation shown in Fig.4 (a) (b) below:

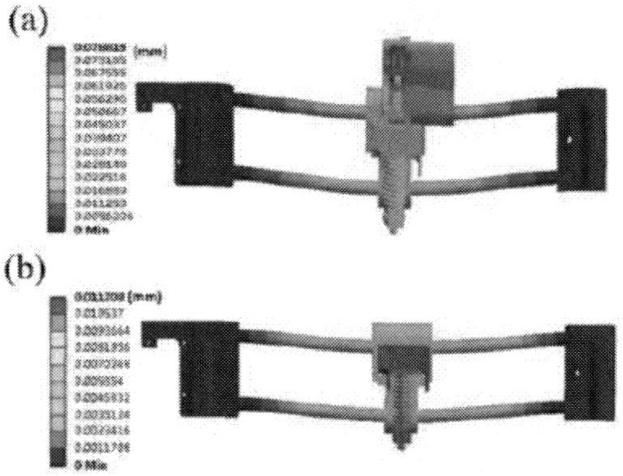

Figure 4. The overall deformation of the nozzle system before separation and after separation

From the static analysis in Fig.4, it can be seen that the largest deformation of the nozzle system before separation is 0.0788mm, and the largest deformation after separation is 0.0117mm. The amount of deformation of the nozzle system after the initial separation compared to before separation has greatly changed.

In view of the improved effect of the statics analysis of the separated nozzle, the design of the nozzle part is optimized. During the working process of the 3D printer, the temperature of the melted PLA is between 180-210 degrees centigrade, and the cooling fan is located on both sides of the heating nozzle. The temperature on the surface of the nozzle is sampled with an infrared thermometer, and the temperature is measured at 47.6 degrees near the fixed end of the nozzle. From the finite element analysis of the whole machine on the 3D printer above, it is known that the fixed part of the material with PLA does not show a larger amount of deformation. Combining the properties of PLA materials, it can be seen that at the working conditions of the 3D printer, the PLA does not show any signs of distortion and melting as a fixed material at the bottom of the nozzle heating system. In summary, PLA meets the requirements of designing nozzle-aided fixed brackets in terms of mechanical strength and temperature resistance.

The model of the heating nozzle is shown in Fig. 5 (a). According to the shape of the nozzle, when the auxiliary fixed bracket is designed, the surface profile of the auxiliary fixed bracket is should matched with the surface of the heating nozzle tail to prevent the heating nozzle from moving during the printer's work, which may lead to the offset of the printed material. In addition, during the design process, the designed auxiliary fixed bracket should satisfy the symmetrical distribution. According to the literature data, it is found that the symmetrical mechanical structure could help to improve the natural frequency of the structure itself. A well designed section of the fixed bracket for the heating part of the nozzle is shown in Fig.5 (b).

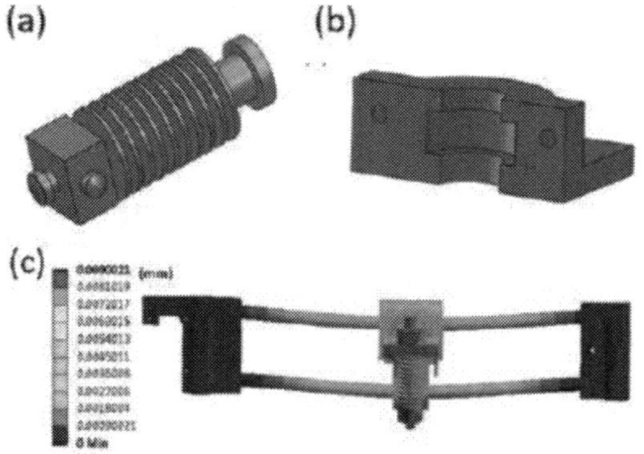

Figure 5. (a)Printer head model and (b)fixed bracket profile And (c)the static analysis of the 3D print nozzle system after the bracket is fixed

The designed fixed bracket and the heating nozzle part are assembled, and the static mechanical analysis of the finite element is carried out. The analysis results are shown in Fig .5(c).

After the installation of the designed fixed bracket, it can be seen from the Fig.5(c) that the maximum deformation of the nozzle system is 0.0090mm, therefore, the maximum deformation of the nozzle system before the separation, after the separation and after the designed fixed bracket installed are shown in table 2.

Table 2. The maximum deformation data table for static analysis

Nozzle system state	Before the separation	After the separation	Designed fixed bracket installed
The maximum deformation (mm)	0.0788	0.0117	0.0090

The frequency analysis of the finite element modal analysis of the nozzle system in the three cases is shown in Fig.6

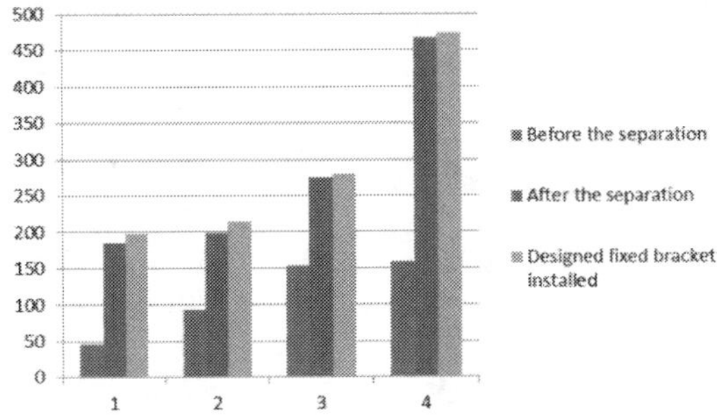

Figure 6. Frequency contrast diagram of modal analysis in 3 cases

From the analysis results, it can be seen that after the designed fixed bracket installed, the overall deformation of the nozzle system obtained is smaller than before separation and after separation, the natural frequency of each order is also greatly improved compared to the before and after separation.

5.2 Influence of changing level shaft bearing properties on the nozzle system

The level shaft bearing plays an important role in the load-bearing and modeling process of the printer. In order to improve the structural stability of 3D printer, the diameter and material of the level shaft bearing is experimentally explored after the designed fixed bracket installed. When the level shaft bearing material and diameter is changed, the response of the nozzle system in statics and dynamics analysis is discussed.

When analyzing the influence of changing the level shaft bearing material on the stability of the printer nozzle system, the material of level shaft bearing is set up as titanium alloy, copper alloy, bearing steel and material property as shown in table 3.

Table 3. Material properties of level shaft bearing

Type of material	Density(Kg/m^3)	Poisson ratio	Modulus of elasticity (Gpa)
titanium alloy	4620	0.36	96
copper alloy	8300	0.34	110
bearing steel	7810	0.3	208

When comparing the influence of changing the level shaft bearing diameter on the stability of the nozzle system, the diameter is set to 6mm, 8mm, and 10mm.The properties of the nozzle system are set up, and the statics analysis of finite element is carried out.

When changing material properties, the maximum deformation of the nozzle system is shown in the table 4.

Table 4. The maximum deformation of the nozzle system on different material properties

Type of material	titanium alloy	copper alloy	bearing steel
The maximum deformation (mm)	0.0118	0.0126	0.009

When changing the nozzle diameter, the maximum deformation of the nozzle system is shown in the table5.

Table 5. The maximum deformation of the nozzle system on different diameter of level shaft bearings

Diameter of level shaft bearing (mm)	6	8	10
The maximum deformation (mm)	0.0147	0.0090	0.0089

The modal frequencies from first to fourth orders of different material properties and diameters are taken during the dynamic finite element analysis. The results are as follows Fig.7.

In the analysis of different diameters about level shaft bearing, the material is structural steel, and in the analysis of different material properties of level shaft bearing, the diameter is 8mm.

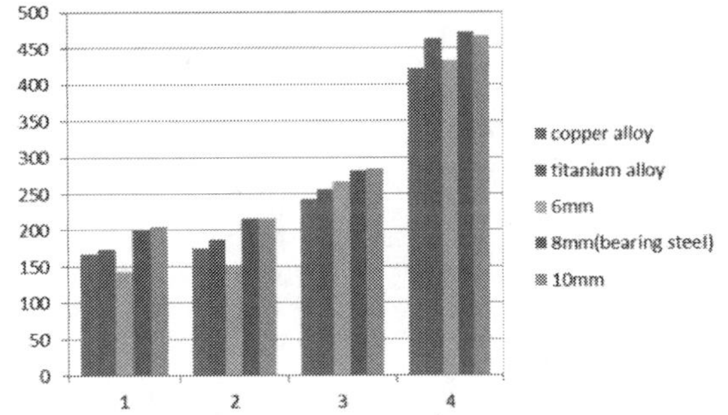

Figure 7. Modal frequency comparison of level shaft bearings under different conditions

It is known from the statics analysis that in the same diameter of the level shaft bearing, the deformation of the material of structural steel is the smallest. At the same time, in the modal analysis of finite element, the modal frequencies from first to fourth orders is higher than other materials. When the material of the level shaft bearing is the same and the diameter is changed, the deformation of nozzle system shows a decreasing trend with the increase of the diameter. There is little difference in the static deformation of the level shaft bearing diameter between 8mm and 10mm, but in the finite element dynamic analysis, modal frequency presents an increasing trend. Meanwhile, when the diameter of the level shaft bearing is from 8mm to 10mm, the increment of the modal frequency is not much, and at the four order modal frequency, the diameter of the level shaft bearing of 8mm is greater than the diameter of 10mm.

It can be seen from the analysis of the conclusion that is no clear relationship between static deformation and the single material property of the level shaft bearing. From the single density, Poisson's ratio, elastic modulus of materials is difficult to judge the final applicability.

For the diameter of the level shaft bearing, when the other conditions are constant, the diameter of the optical axis is not very suitable to increase endlessly, normally, the increase of level shaft bearing diameter helps to improve the bearing capacity and modal frequency of the nozzle system. At that time, it reaches a certain limit value, the effect of the level shaft bearing diameter would not be very obvious, therefore, in the flowing study, the material of the optical axis is set up to the structure steel, and the diameter is 8mm.

 https://doi.org/10.1051/matecconf/201925702004

6 The overall structural optimization of the 3D printer

6.1 Theoretical optimization design of mechanical structure

From the overall structure static finite element analysis of 3D printer, it can be seen from the Fig2.(b) that the deformation of the platform is asymmetrical. The main reason is that the supporting bearing under the forming platform is asymmetrical, and it a fixed linear bearing needs to be added on the larger deformation side of the platform to ensure a good symmetry of the forming platform. In addition, before the optimization of the 3D printer, it can be obtained from the first to fourth modal frequencies that the two side columns and beams of the 3D printer have a more serious phenomenon of vibration. And in the statics analysis Fig2.(a), it is also found that the deformation of 3D printer`s column is still about 0.01mm to 0.03mm.

In order to solve these problems, the columns on both sides of the printer are symmetrically fixed by 2020 aluminum alloy material, finally, the whole frame of the 3D printer is square. On this basis, after the designed fixed bracket installed, the nozzle system is fixed on the whole frame of the printer.

The finite element analysis of the statics and dynamics of the overall 3D printer structure is carried out, and the deformation of the overall structure statics analysis of the optimized printer is shown in Fig.8.

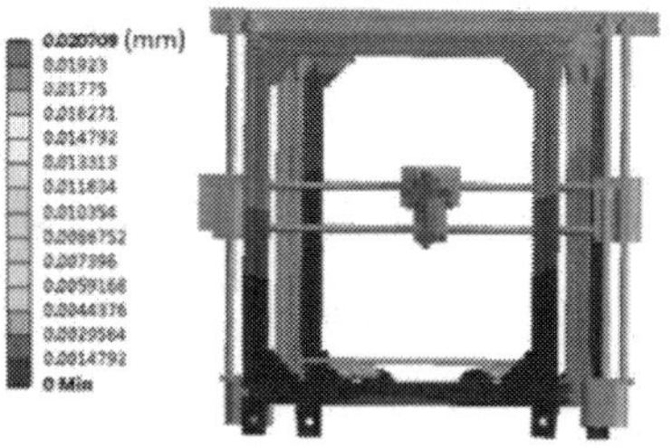

Figure 8. Improved deformation of the overall structure of the 3D printer

It can be seen from the static analysis of the overall structure that the largest deformation of the optimized 3D printer is 0.0207mm, located at the nozzle system. Compared with the previous deformation of 0.117mm, the statics deformation of the optimized 3D printer has been better improved. The overall structure modal analysis of optimized printer is shown in Figure 9.

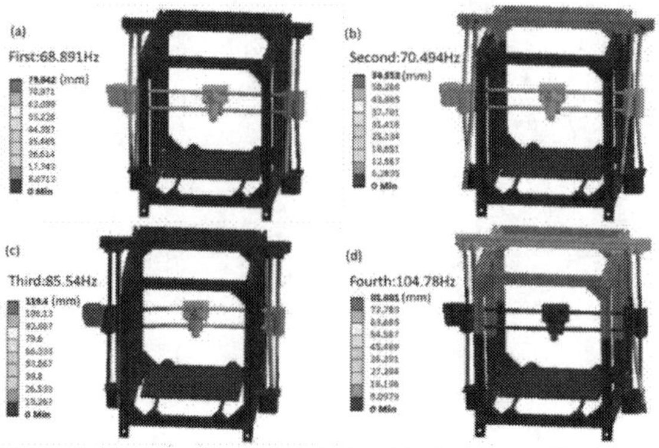

Figure 9. The overall modal analysis of the improved 3D printer

In the result of modal analysis, the 3D printer`s natural frequency of each order is higher than that of the original structure. When the vibration modes are observed at each order of nature frequency, at a lower frequency, there is no serious vibration in the column and beam of the improved 3D printer structure, and only at the second and fourth order natural frequency, there is a small vibration at the beam. In addition, the forming comparison between the before optimized and after optimized of 3D printers is carried out, it is showed in Figure 10.

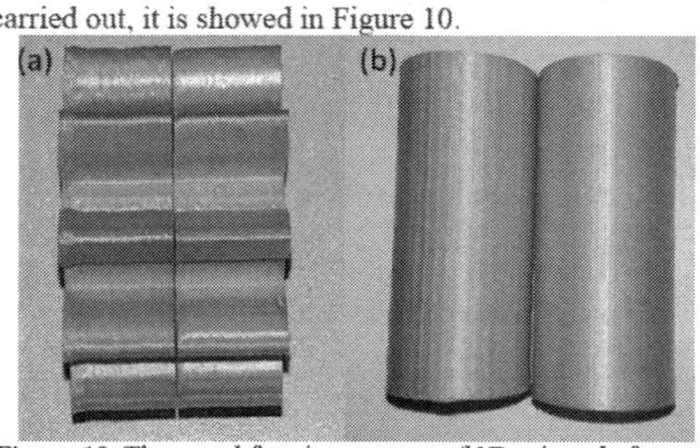

Figure 10. The actual forming contrast of 3D printer before and after improvement

The formed object on the right side of the two pictures is formed by the optimized 3D printer, and the other conditions are the same in addition to the subtle difference in the color of the molding material. As can be seen from the surface accuracy of the formed object, the smoothness of the surface is greatly improved compared to the forming precision of the 3D printer before the optimization.

7 Summary

From the overall structure of the 3D printer, the stiffness of the original column auxiliary fixing device is relatively poor, and there is more obvious vibration phenomenon in the finite element analysis. After the optimization of the 3D printer structure, its overall frame is square, and the natural frequency of each order is greatly improved, the mode of vibration is also more stable, and the maximum deformation of the 3D printer overall structure is reduced from the 0.117mm to the 0.021mm, the surface plating of the formed object also has a better improvement.

As far as the nozzle system of the 3D printer, its quality

should be appropriately reduced during the design process, and it has a great effect on reducing the load of the optical axis and improving the precision. The diameter of the moving level shaft bearing used should be appropriate, when the diameter of the level shaft bearing reaches a certain limit, with the increase of diameter, the dynamic parameters and statics parameters of the nozzle system will not be changed significantly. In addition, when the 3D printer is designed or the related parts are optimized, as far as possible to ensure the structural symmetry, it is important to improve the stability and the forming precision of the 3D printer structure. Moreover, the introduction of vibration in the 3D printer forming process should be reduced. In a word, after the structure of the printer optimized, the 3D printer has achieved good results in the finite element analysis and the actual modeling.

Acknowledgment

The project is supported by major scientific and technological projects in Zhejiang Province, No. 2015C01035, Bidding project of the science and Technology Department of Zhejiang Province No.2017C01019, China. I am very grateful to Professor Tingchun Shi for his guidance in writing this essay. And thanks to Yuan Fan, Nannan Sun and Liangliang Lu in literature inquiries

References

1. B. Berman, "3-D printing: The new industrial revolution," Business Horizons,vol. 55,no. 2, pp.155-162,2012.
2. G. Cesaretti, E. Dini, X. D. Kestelier, V. Collaand L. Pambaguian, "Building components for an outpost on the Lunar soil by means of a novel 3D printing technology," ACTA ASTRONAUT,vol. 93,no. 1, pp.430-450,2014.
3. B. T. Wittbrodt, A. G. Gloverand J. Laureto et al., "Life-cycle economic analysis of distributed manufacturing with open-source 3-D printers," MECHATRONICS,vol. 23,no. 6, pp.713-726,2013.
4. B. M. Tymrak, M. Kreigerand J. M. Pearce, "Mechanical properties of components fabricated with open-source 3-D printers under realistic environmental conditions," MATER DESIGN,vol. 58,no. 6, pp.242-246,2014.
5. M. Merzoug, M. Mazari, L. Berrahaland A. Imad, "Parametric studies of the process of friction spot stir welding of aluminium 6060-T5 alloys," MATER DESIGN,vol. 31,no. 6, pp.3023-3028,2010.
6. C. X. Yue, L. W. Zhangand S. L. Liao et al., "Research on the Dynamic Recrystallization Behavior of GCr15 Steel," Materials Science & Engineering A,vol. 499,no. 1–2, pp.177-181,2009.
7. A. K. Bledzki, A. Jaszkiewiczand D. Scherzer, "Mechanical properties of PLA composites with man-made cellulose and abaca fibres," Composites Part A Applied Science & Manufacturing,vol. 40,no. 4, pp.404-412,2009.
8. T. T. Ngoc, G. R. Liu, H. Nguyen Xuanand T. Nguyen Thoi, "An edge - based smoothed finite element method for primal–dual shakedown analysis of structures," INT J NUMER METH ENG,vol. 82,no. 7, pp.917-938,2010.
9. S. Shojaee, N. Valizadeh, E. Izadpanah, T. Buiand T. V. Vu, "Free vibration and buckling analysis of laminated composite plates using the NURBS-based isogeometric finite element method," COMPOS STRUCT,vol. 94,no. 5, pp.1677-1693,2012.
10. S. Nikbakht, S. J. Salamiand M. Shakeri, "Three Dimensional Analysis of Functionally Graded Plates Up to Yielding, Using Full Layer-wise Finite Element Method," COMPOS STRUCT,vol. 182,2017.
11. T. Csanádi, D. Némethand F. Lofaj, "Mechanical Properties of Hard W-C Coating on Steel Substrate Deduced from Nanoindentation and Finite Element Modeling," EXP MECH,vol., pp.1-13,2016.
12. N. Valizadeh, S. Natarajanand O. A. Gonzalez-Estrada et al., "NURBS-based finite element analysis of functionally graded plates: Static bending, vibration, buckling and flutter," COMPOS STRUCT,vol. 99,no. 5, pp.309-326,2013.
13. C. C. Hong, C. L. Changand C. Y. Lin, "Static structural analysis of great five-axis turning–milling complex CNC machine," Engineering Science & Technology An International Journal,vol. 19,no. 4, pp.1971-1984,2016.
14. Z. Silvayeh, R. Vallantand C. Sommitsch et al., "Influence of Filler Alloy Composition and Process Parameters on the Intermetallic Layer Thickness in Single-Sided Cold Metal Transfer Welding of Aluminum-Steel Blanks," Metallurgical & Materials Transactions A,vol. 48,no. 11, pp.5376-5386,2017.
15. D. Moensand M. Hanss, "Non-probabilistic finite element analysis for parametric uncertainty treatment in applied mechanics: Recent advances," Finite Elements in Analysis & Design,vol. 47,no. 1, pp.4-16,2011.

Online Detection and Study of Cracks in Grinding Hardening

Liang Ping [1], Cheng Wei [1], Zhao Yulong [1]

[1] *Huaiyin Institute of Technology, Huai'an 223003, China*

Abstract. Grind-Hardening technology is a new technology for cutting that combines grinding theory with quenching theory, and the heat produced during the grinding process has a great effect on the surface quality of the parts. This paper take advantage of method of acoustic emission testing to test the grinding hardened parts by online or scene. The wavelet packet is used to denoise the signal, then the denoising signal is analyzed to judge those grinding cracks or defects that may appear on the surface of the parts, and besides, this method is easy to do.

1 Introduction

Grind-Hardening is a new technology of cutting that combines grinding theory with quenching theory , and the technology is a integration of grinding process and surface quenching , which can effectively shorten production cycle, improve production efficiency, decrease manufacturing cost ,reduce the input of heat treatment equipment and personnel , as well as the pollution from heat treatment emissions to the environment , has prominent economic value and social benefits . In the grinding process, on the surface layer of the part may occur cracks or micro cracks if the grinding parameters are not chosen properly. When the temperature of grinding zone exceeds a certain threshold, the residual stress on the surface of the part may exceed the strength limit of the material, resulting in grinding cracks at the instantaneous cooling of the grinding liquid [1].

2 Acoustic emission testing of grinding crack

The appearance of grinding cracks in the surface layer has a great influence on the performance of the part in the grinding process, so online or scene testing has profound significance for timely detect crack generation or cracks. This paper adopts acoustic emission to test the grinding crack of parts.

Acoustic emission refers to the phenomenon that local material emits transient elastic wave due to the rapid release of energy. Acoustic emission testing is a very effective method to test the trend change of material under stress. The basic principle of acoustic emission testing is shown in figure 1[2]:

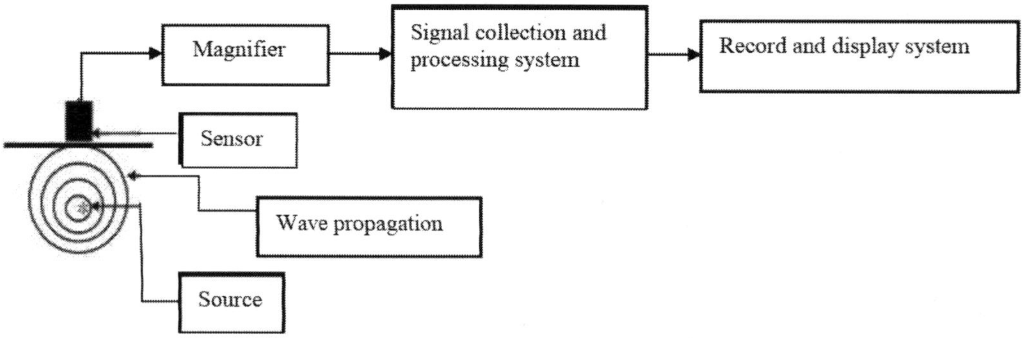

Figure 1. Acoustic emission testing principle diagram

The elastic waves emitted from the acoustic emission source propagate in the material ,finally get to the surface of the material , causing surface vibration signal that can be detected by acoustic emission sensor , as well as these sensors convert the mechanical vibration of the material into electrical signal and then magnify them with magnifier . Signals are collected, stored and processed in real-time by the signal collection cards and the corresponding collection software. Changes of internal stress in the solid material can produce acoustic emission signal , and there are many factors may cause internal stress changes in the process of material processing , such as dislocation movement , twinning , crack initiation and expansion , fracture , thermal expansion and cold contraction ,load change ,etc. Research shows that many acoustic emission source produce transient elastic wave which can be received by taking advantage of special electronic instruments after a series of reflections and refraction to the surface of the material, then these electrical signals are converted into digital signals by a series of means to analyze them for useful information. Acoustic emission is a dynamic nondestructive testing method, which can monitor the object in real-time and high detection sensitivity. Moreover ,acoustic emission detection is not limited by material for almost all material have acoustic emission characteristic , and the size , geometry , working environment and other factors of the object have no effect on the testing[3,4].

For grinding crack detection, we choose digital acoustic emission detector of Beijing soundwel SAEU2S to collect signals, by using the signal analysis function of MATLB wavelet packet to find the singularity in the signal to determine the position of the crack[5].

3 Acoustic emission signal analysis of grinding crack

The main idea of denoising based on wavelet packet transform is to use multiscale characteristic of wavelet analysis. Firstly, wavelet transform is performed on noisy images. Secondly, the wavelet coefficients are processed by thresholding to get new wavelet coefficients. Lastly, the wavelet coefficients are processed with inverse transformation to get denoised images, realizing the recovery of the images. At present, many new image denoising algorithms have been developed by combining wavelet transform with traditional image denoising methods, they absorb advantages of both, and thus acquire a bright future for improving the denoising effect.

4 Acoustic emission acquire signal

The acoustic emission signal is collected by selecting those parts with cracks and no cracks on the surface after grinding hardening, then signal data is exported to MATLAB software to denoise the above signals by using MATLAB wavelet packet. Invoking the wavelet analysis toolbox, as shown in figure 2:

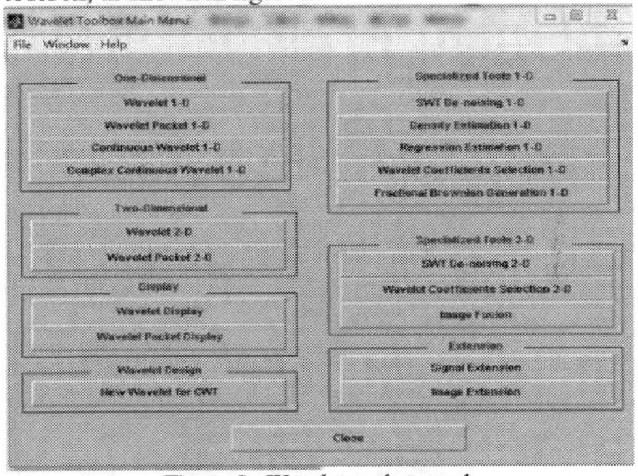

Figure 2. Wavelet packet panel

Selecting wavelet1-D,then import signal to be denoised, And wavelet is decomposed into a wave with 5 layers of db4 as wavelet base, as shown in figure 3:

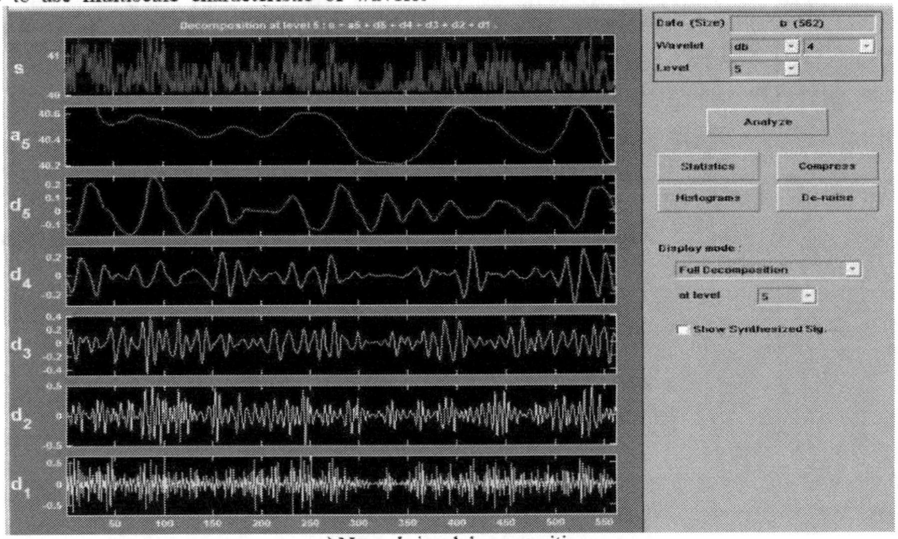

a) Normal signal decomposition

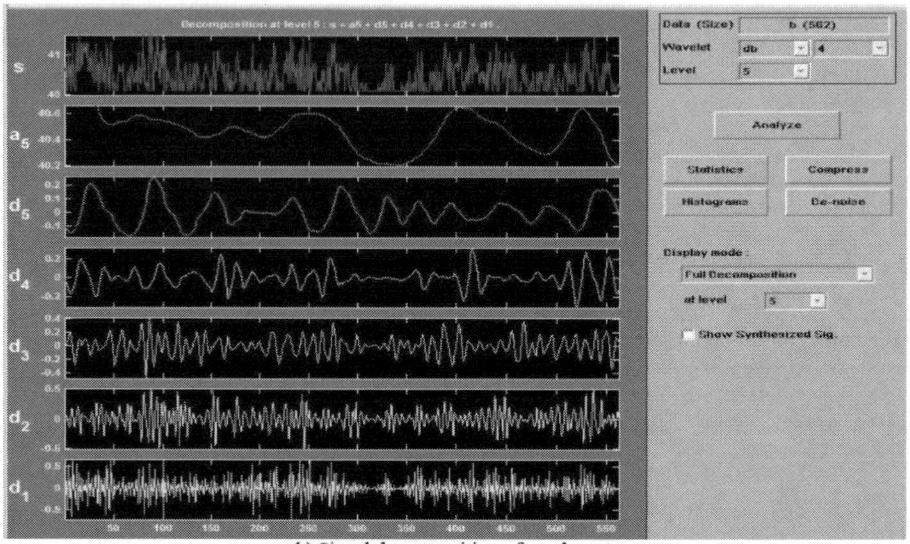

b) Signal decomposition of crack part

Figure 3. Decomposition of normal signal and cracked signal

Opening the denoised image and dividing it into 5 layers of db4 wavelet to analyze it, as shown in figure 4:

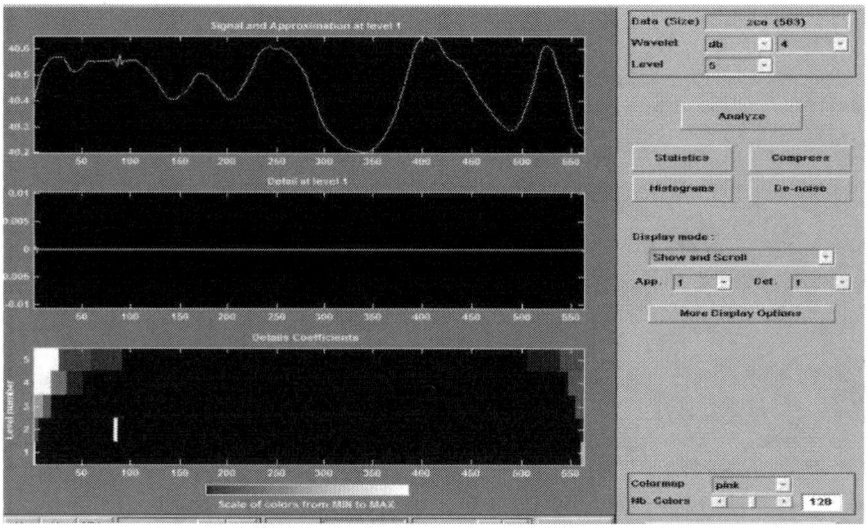

a) Denoising signal and wavelet transform diagram of normal signal

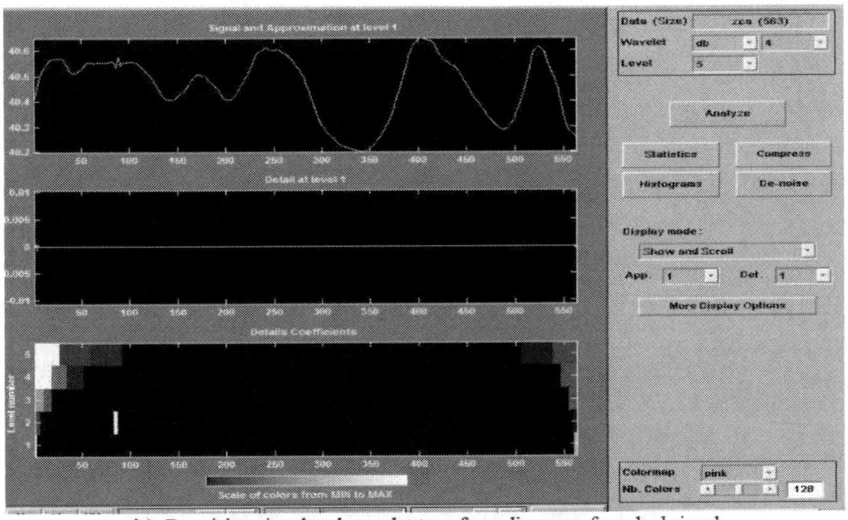

b) Denoising signal and wavelet transform diagram of cracked signal
Figure 4. Denoising signal and wavelet transform

The coefficients are present in the b diagram from the first to the fifth floors and the wavelet coefficients have an inverted cone area at X=+86.694 and X=+391.62,which can be seen clearly in the diagram of the continuous wavelet transform coefficients. As a result, it is inferred that there is a mutation point in this area, so it can be concluded that crack defects exist in this region when grinding parts do not exist other defects or other defects have been detected.

5 Conclusion

The crack defects on the surface of the parts can be detected directly by comparing the wavelet data on the surface of normal parts and those parts with cracks on the surface after grinding hardening. Further, the location and size of cracks can be analyzed to provide technical support for active control surface quality of grinding hardened, development of integrated manufacturing by grind-hardening technology and Engineering application of automation technology.

References

1. Cheng wei, Wang guicheng. Automatic Control Technology of Grinding Temperature in Grinding Hardening Process. Advanced Materials Research, Vol.381:48-51.
2. LONG Xiao-jiang. Acoustic emission monitoring and evaluation for rolled steel damage under different tensile rates [J] . Journal of Vibration and Shock, 2017, 36(7):219-225
3. LUO Zhi-gao, et al. State identification of a punching die based on characteristic analysis of acoustic emission signal [J] . Journal of Vibration and Shock, 2009, 28(3):186-190.
4. SHEN Gong-tian, et al. Parameter analysis of acoustic emission signals [J] . Nondestructive Testing(China), 2002, 24(2): 72-77.
5. WANG Wen-you. Acoustic Emission Testing Technique in Leak Detection of Pressure Pipeline[J]. Nondestructive Testing (China), 2016, 38(2): 22-24.

Designing and Measuring the Hot-Embossing Machine with Hybrid Positioning Functions

Cheng-Hsien Wu, Cheng-Hao Chiu

Department of Mold & Die Engineering, National Kaohsiung University of Science and Technology, Kaohsiung County

Abstract. The demands of the ability of industrial automotive production are rising. The precision positioning is the most important among the manufacturing and measuring. Molds might result in making equipment deviations via poor accuracies on mass production because of the poor design and front processing. It influences the qualities of products, or damages the mold themselves. Based on the Hot-Embossing process, the study designed a set including the Hot-Embossing machine with platform for positioning system holding the mold. The positioning and adjusting function, with X, Y, C axis, separates the Hot-Embossing machine from the traditional one.
Taken smart phone case as the experimental designed concept and verified the pressing molds with the final thickness with the before and after correction for lengths, wide, and the error values of horizontal symmetry to investigate the accuracies of positioning capability. Then, verifying the Hot-Embossing machine from reconstruction of the product.

1 Introduction

Most of the products which made from high polymer materials are in need of molds to be shaped. After designing and machining, however, the molds allowed the error accuracy in the components of parts. The problem of inaccurate position would only occur when tolerances does in assembling. Take producing dual -sided microstructure as an example; the reducing of the product functionalities might result in dual-sided positioning error via assembly tolerances[1]. As shown in figure 1, the dual-sided micro lens array, the original design is the double-sided aliquots arrangement. After operating, however, the excessive concentration of light diffused distribution occurs because the dislocation of dual-sided structure via position error as shown in figure 2, the light distribution of the dislocation of dual-sided micro lens array.

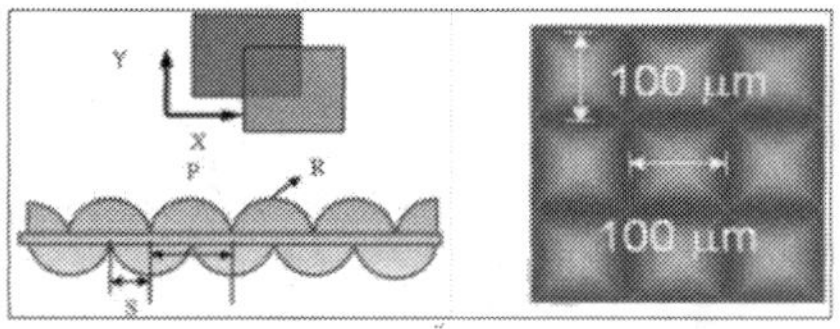

Figure 1. Sketch map of dual-sided micro lens array

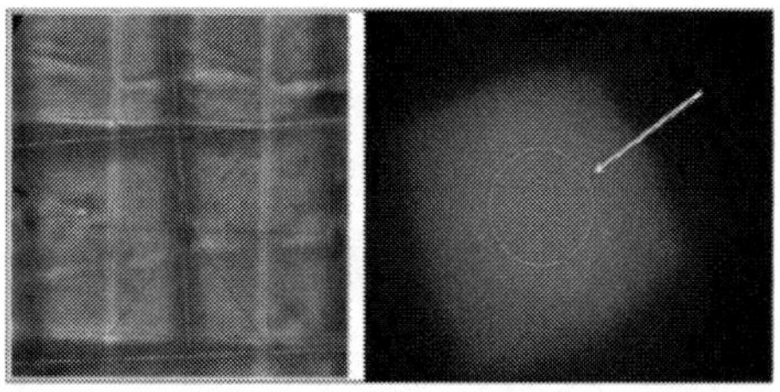

Figure 2. Light distribution of the dislocation of dual-sided micro lens array

To solve the problem above, it has to disassemble molds, re-position, and reassemble; or vary molds. The process would not be just one time. This reduces the efficiency, increases cost, takes times, and influences capacity. Therefore, designing one loader which could adjust immediately after testing can solve the problem efficiently. There are many patents and literatures on studying the function of positioning platform; and utilizing external loader for Hot-Embossing process. It is because the lack attentions of positioning ability on Hot-Embossing machine in industry. Thus, the study designed a micro Hot-Embossing machine based on Hot-Embossing and to economize space, disassemble and assemble easily, simplifies mechanical structure, operate easily, and include precision positioning capability. Taking thin-shell molding as the experience demonstrates the positioning accuracy and reconstruction for the relative study in the future.

2 Experiment

2.1 Design concept and demand

A precise Hot-Embossing machine, shown in Fig. 3, is specially designed for the accurate replication of optical elements. The Hot-Embossing is calibrated and adjusted to improve its alignment and parallelism. A loaded cell was applied to monitor the pressing force. The measured force was also used to control the embossing at a specified pressing force. Four cartridge heaters were applied to heat the plates. Two heaters were installed in the upper plate and the other two were installed in the lower plate. These four cartridge heaters were controlled by two thermocouples installed in the two plates. The temperatures of the two plates were averaged and made use of controlling the mold temperature. Another thermocouple was installed on top of the polymer substrate to monitor the material temperature. The Hot-Embossing was regularly calibrated and adjusted to improve its alignment and parallelism.

2.2 Hot-embossing machine and X-Y horizontal positioning and adjusting platform

The basic demands for making a Hot-Embossing machine are dynamics, structure and function, and control. According to the theory above, setting the Hot-Embossing machine with hybrid position functions in the study as one combines the features of relative process and the lacks of tradition one. The basic structure chart in figure 3 is the technical architecture for the study. The sketch map of perspective view of the machine and the assembly horizontal slider are shown in figure 4 and 5.

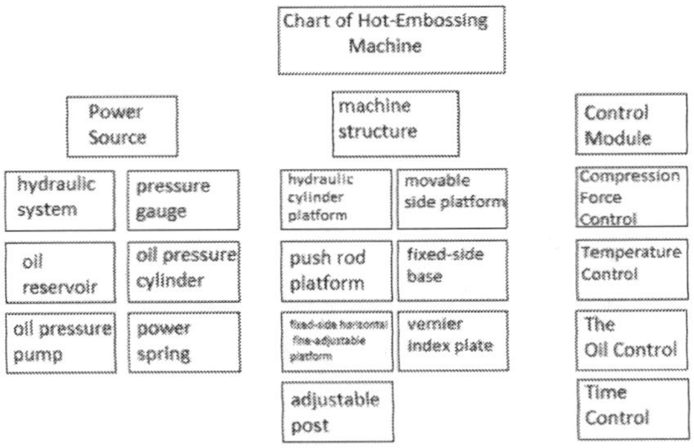

Figure 3. Technical architecture of Hot-Embossing machine with the hybrid positioning function

Figure 4. Sketch map of Hot-Embossing machine with hybrid positioning function

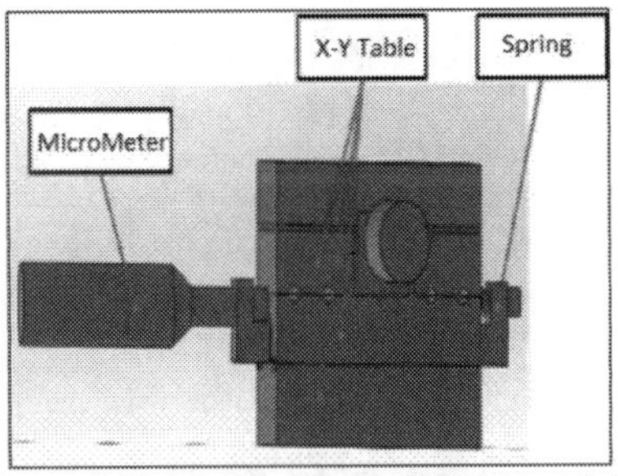

Figure 5. Sketch map of assembly horizontal slider

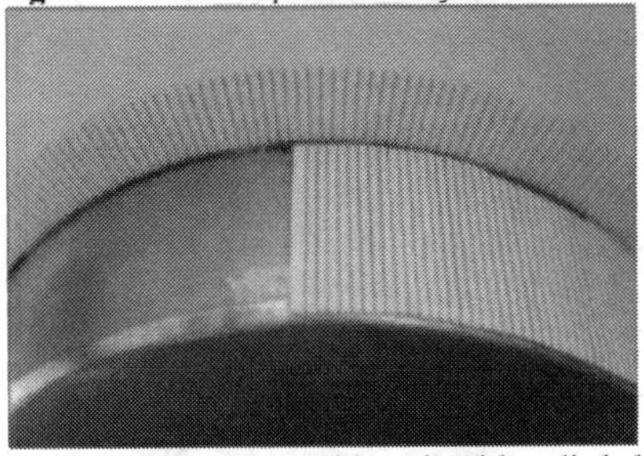

Figure 6. Analytical scale table and real installed photo

2.3 Rotary positioning platform

As figure 6, the rotary spinner platform is combined with main scale fixed tray and venier scale turntable. Using steel S50C supports the high load. According to the universal protractor theory, divided the main scale fixed tray from 1 degree of each section, and the venier scale turntable into 30 equal parts with 29 degrees of main scale fixed tray in each part, which means that each venier scale section is 0.97 degree. While the first engraving line of venier scale aim at the 0 degree of main scale fixed tray, both of them will have a gap in 0.03 degree. The resolution of rotary positioning platform will be 2' after unit transforming. To lighter the weight of machine, the study gear the tray and turntable via movable screw with bearing fasten through guide pillars and side board.

https://doi.org/10.1051/matecconf/201925702006

2.4 Experience procedures

The design processes of the study are as follow:

I. Assembling and designing machine
II. Proofing stabilities of forming machine
III. Proofing stabilities of positioning system
IV. Proofing actual product

2.5 Design of experimental molds

In order to manufacturing a shell product with height, depth, and width; the experience utilized FDAC steel to devise a closed Hot-Embossing mold. it utilize its measuring values of the side thickness to revise the positioning system of Hot-Embossing machine and to devise the adjusting capability and function of it.
As shown in figure 7, the under mold can separate into two shelf. First, the lower fixed plate is to fixed the under molds and the main structure of positioning system. The second shelf is the down template, it is made by high precision machining to combine the insert and template. The three penetrating horizontal holes are used to set up electrical bars (ø10mm) and temperature sensors (ø5mm), also the cooling channel and waterways connector thread (ø8mm).

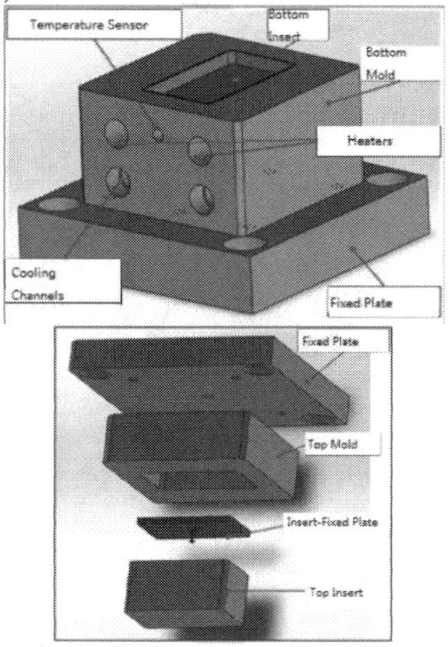

Figure 7. Sketch map of down molds (left) and up molds (right)

3 Results

Hot-Embossing machine with hybrid positioning function developed according to the theories shown as figure 8 mentioned above. It is conducted experimental test respectively. The test will focus on excluding the abilities to adjust horizontal axis X-Y, rotation angle axis C, and the reproduction, also the disadvantages of product and forming characteristics.

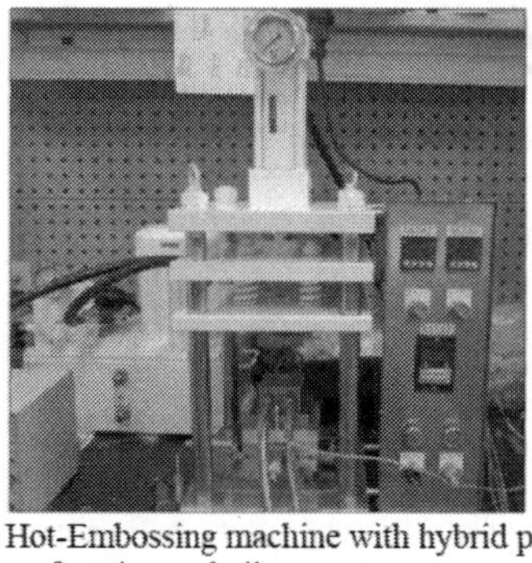

Figure 8. Hot-Embossing machine with hybrid positioning function and oil pressure system

3.1 Steps and parameters of testing positioning technique

The steps of the positioning technique of Hot-Embossing machine are as follow:

I. Setting the up molds and down molds respectively on the rotary platform and horizontal platform of Hot-Embossing machine with hybrid positioning function.
II. Taking the polymer materials properly via $\rho = M/\square$, density volume quality and quantity formulation, and putting into the down insert.
III. Making by Hot-Embossing process: preheat, Hot-Embossing, cooling, open
IV. Utilizing OM measures preliminarily to confirm if the product is in the right position via optical electron microscope (OEM).
V. Utilizing OM measures to confirm it again via OEM VHX-1000 and recording the data.

The polymer materials, PMMA, which was used in the study, were produced by CHIMEI Corporation. Parameters, shown in table-1, were set according to Yao[2] and Heckele[3]. Demonstrating angles adjust; there are 6 groups of data in total, adjusting 12' in each time. Then, demonstrating horizontal adjusts; there are 20 groups of data, adjusting axis X and Y 9 times separately based on 30 μm in each adjustment. Demonstrating reproduction of product in the third part, the 12 relative positions on the four sides of product surveyed the relation of thickness, as shown in figure 9.

Table 1. Experimental parameters in testing positioning technique

	Temperature (℃)	Compression (kgf)	Time(s)
Pre-Embossing	160	700	60
Hot-Embossing	160	2900	900
Cooling	65	2900	60

Figure 9. Sketch map of measured position of reconstruction verification

3.2 Result of testing positioning technique

Adjusting process of product measurement is shown in Table 2.

I. By OEC VHX-1000, the angle positioning test result is as follow:
II. First Hot-Embossing testing of C-axis, the angle value is 89.65 degrees. The value was 89.90 degrees after first adjusting. Following is 90.09 degrees. The third is 90.26 degrees. Next is 90.49 degrees. It is 90.71 degrees after the fifth adjusting.
III. By OEC VHX-1000, the X axis positioning test [4]result is as follow:
IV. First Hot-Embossing testing of X-axis, the value is 690.13 μm. After first adjusting, the value was 719.85 μm. Following is 747.44 μm. The third is 779.29 μm. Next is 809.03 μm. It is 838.73 μm after the fifth adjusting. Then, it is 870.75 μm at the sixth adjusting. Following is the 900.43 μm. Next is 930.10 μm. Final, the ninth adjusting is 959.80 μm.
V. By OEC VHX-1000, the Y axis positioning test result is as follow:
VI. First Hot-Embossing testing of Y axis, the value is 747.44 μm. After first adjusting, the value was 775.03 μm. Following is 804.78 μm. The third is 834.51 μm. Next is 864.31 μm. It is 894.03 μm after the fifth adjusting. Then, it is 925.88 μm at the sixth adjusting. Following is the 955.63 μm. Next is 985.36 μm. The ninth adjusting, the value comes to 1014.97 μm. Table 2 is the analysis of this result.

Reiteration of the product is verified by the thickness of mold around edges which settle on 800 μm at one edge, 1600 μm at two edges. The marked position in figure 9 gauged the relative thickness of production. The summation is shown in Table 3.

Table 2. Analysis table of result of adjusting process of product measurement

	X-axis (μm)	Alignment Error with Last time	Y-axis (μm)	Alignment Error with Last time	C-axis	Alignment Error with Last time
Pre-Pos.	690.13		747.44		89.65°	
1	719.85	29.72	775.03	27.59	89.9°	0.25°
2	747.44	27.59	804.78	29.75	90.09°	0.19°
3	779.29	31.85	834.51	29.73	90.26°	0.17°
4	809.03	29.74	864.31	29.8	90.49°	0.23°
5	838.73	29.7	894.03	29.72	90.71°	0.22°
6	870.75	32.02	925.88	31.85		
7	900.43	29.68	955.63	29.75		
8	930.1	29.67	985.36	29.73		
9	959.8	29.7	1014.97	29.61		
	AVG	29.96	AVG	29.72	AVG	0.21°
	Adjustment Accuracy	99.80%	Adjustment Accuracy	99.00%	Adjustment Accuracy	105%
	Alignment Error	-0.20%	Alignment Error	-1%	Alignment Error	+5%

 https://doi.org/10.1051/matecconf/201925702006

Table 3. Analysis table of product reconstructions

Y-axis			
Point.1	Point.2	Total	Alignment Error
949.15	662.83	1611.98	11.98
Point.3	Point.4	Total	Alignment Error
955.71	671	1626.71	26.71
Point.5	Point.6	Total	Alignment Error
951.27	662.51	1613.78	13.78
X-axis			
Point.7	Point.8	Total	Alignment Error
770.81	840.86	1611.67	11.67
Point.9	Point.10	Total	Alignment Error
768.66	840.86	1609.52	9.52
Point.11	Point.12	Total	Alignment Error
772.91	838.74	1611.65	11.65
Design of Sidewall Thickness(T) is 1600μm for Mold-insert		Average Alignment Error	14.21

3.3 Steps of applying positioning technique

The prospect inspired by cell phone shell, we want to design a shell product with even thickness in four edges. It is because the uneven symmetry of shell might result in the deviations of angles and horizontal positioning. The uneven symmetry might occur the situations of defective assembly or unable to assemble.

The steps of applying positioning technique are as follow:

I. Settling the mold on the Hot-Embossing machine that develops in this prospect.
II. Beginning the first Hot-Embossing, and measuring the thickness and the angular deviation of product
III. Doing the one-off positioning on the measured angular deviation. After it, the angle should be positioned, and do the Hot-Embossing again.
IV. Surveying the thickness deviation on four edges after second Hot-Embossing, and doing the one-off horizontal axis positioning on the deviation. After one-off adjustment, the thickness should be right and could begin the third Hot-Embossing.

Gauging the thickness of four edges after one-off adjustment and seeing if fill the conclusion of positioning technique mention above.

3.4 Results of applying positioning technique

As shown in figure 10(a), the initial Hot-Embossing angular measured value is 89.65° which has 0.35° error compared with the expectation 90°. The error would be 21' after unit transforming. Since the minimum amounts that are able to adjusted is smaller than 2', the designing resolution of positioning platform, thus the default adjusting value set as 22'. It is shown in figure 11(b); gauging the value after one adjusting is 90.03°, which actually move 22.78' after unit transforming. It has 0.78' error, which is 0.013° error after unit transforming.

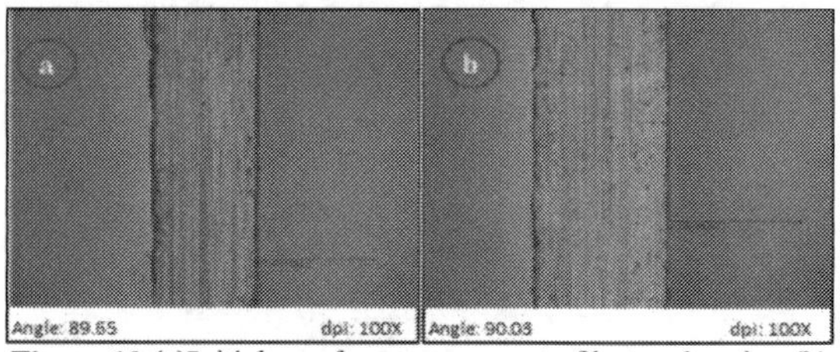

Figure 10.(a)Initial angular measurement of hot-embossing (b) Angular measurement after one adjusting

As shown in figure 11(a) and (b), the thickness of shell on X axis is 643.47 μm after angular adjusting, Y axis is 953.40 μm. Hence, a deflection error 156.53 μm compare with expected value 800 μm on X axis, and a deflection error 153.40 μm compare with expected value 800 μm on Y axis. Since the minimum amounts that are able to adjusted is smaller than 10 μm, the designing resolution of positioning platform, thus the default adjusting value set as 160 μm on X axis and 150 μm pm Y axis. It is shown in figure 11(c) and (d); gauging the X axis directive measurement value after one horizontal adjusting is 804.76 μm, which actually move 161.29 μm after unit transforming. On the other hand, Y axis directive measurement value is 804.76 μm after one horizontal adjusting is 804.76 μm, which actually move 148.64 μm after unit transforming.

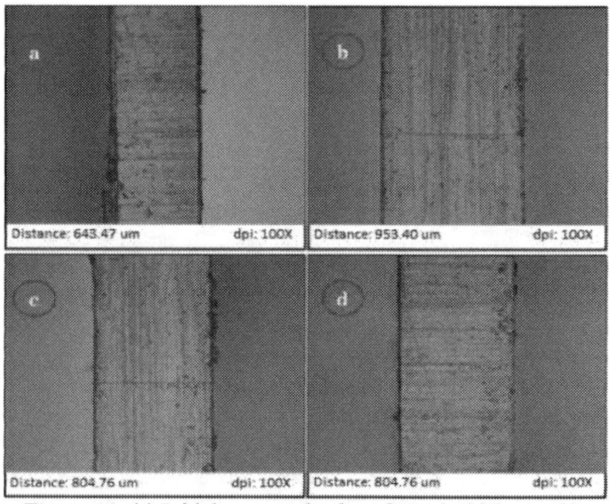

Figure 11. (a) Initial measured value of X axis thickness (b) Initial measured value of Y axis thickness (c) X axis thickness after one adjusting (d) Y axis thickness after one adjusting

3.5 Analyze the influences of hot-embossing pressures to the product

The shell production, which was produced in the experience, was made from acrylic particles. The set is 50mmx30mmx5mm, the thickness of four edges is set as 0.8mm, and the bottom thickness is set as 0.5mm. It is a structure with height, depth, width. Settle a group of parameters of Hot-Embossing pressures to discuss the influences in process of making product from granular

https://doi.org/10.1051/matecconf/201925702006

materials, and discuss the result via surface structures that observed by OEM.

Form 4 is the parameters setting in the experience. Among it, time for preheating is 60 seconds, temperature for preheating is 160℃, and pressures for preheating is 700 kg. Besides, time of keeping pressures is 60 seconds, temperature for keeping pressures is 65℃, and pressures for keeping pressures is 2900kgf. Furthermore, time for Hot-Embossing is 900 seconds, and temperature for Hot-Embossing is 160℃. Main testing parameters of Hot-Embossing were separated into 4 groups.

3.6 Analyze the parameters of hot-embossing pressures

As shown in figure 12, when the Hot-Embossing pressures were set in 1700kgf, the depth and width traces of crystalline phase on the surface are obvious. In addition to crystallizations could not complete welding, the product could not be shaped.

As shown in figure 13, when the Hot-Embossing pressures were set in 2000kgf, the depth and width traces of crystalline phase on the surface were smaller than the first prospect. The welding between crystallization is higher. The product had an embryo of shaping.

As shown in figure 14, when the when the Hot-Embossing pressures were set in 2300kgf, the depth and width traces of crystalline phase on the surface were presented as tiny lines. The ratios of lines were significantly reduced on the product area. The heights of four edges were nearly to the forming altitude.

As shown in figure 15 when the Hot-Embossing pressures were set in 2300kgf, the depth and width traces of crystalline phase on the surface were quite slight. The ratios of lines were unapparent on the product area.

According to the experiences above, choosing granular materials on making a product with height, depth, and width with specific high temperature, thought the high temperature is helpful for crystallizations to be heated and distorted, the deficiency of Hot-Embossing pressures will make the crystallizations too loose to weld, furthermore, inadequate strength will bring defects on the product or unable to shape because the materials are not able to go deep into insert. From the experience, the long sides, which have short distance with the mid, had good flow, on the other hand, the short sides, which were far away from mid, had clear drop.

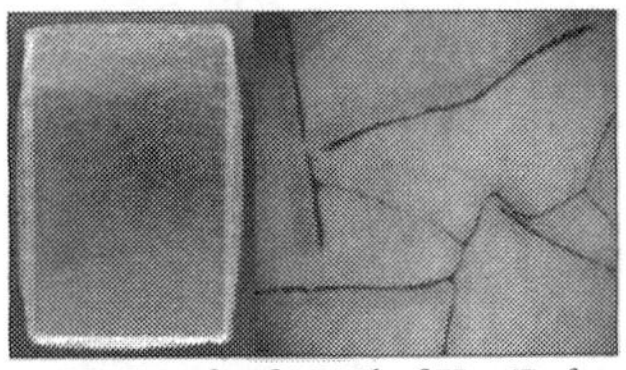

Figure 12. Result of 1700kgf Hot-Embossing

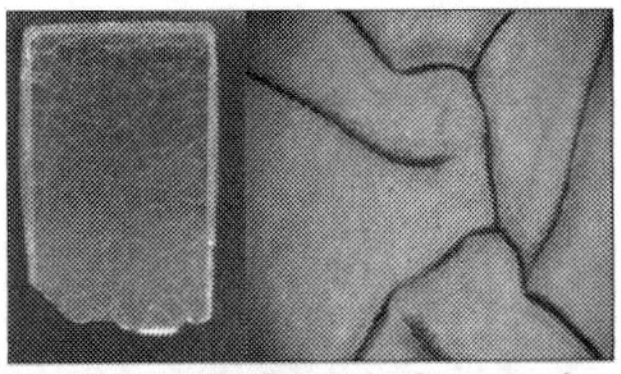

Figure 13. Result of 2000kgf Hot-Embossing

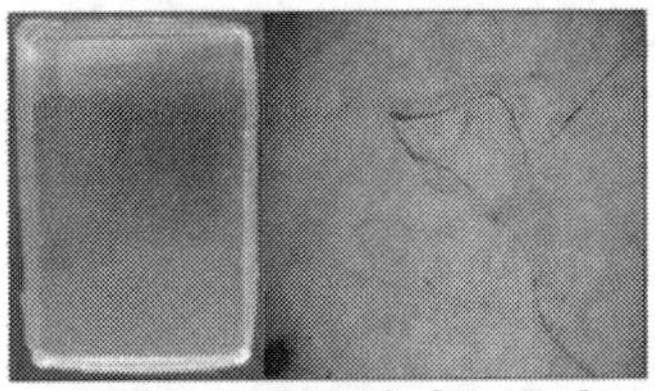

Figure 14. Result of 2300kgf Hot-Embossing

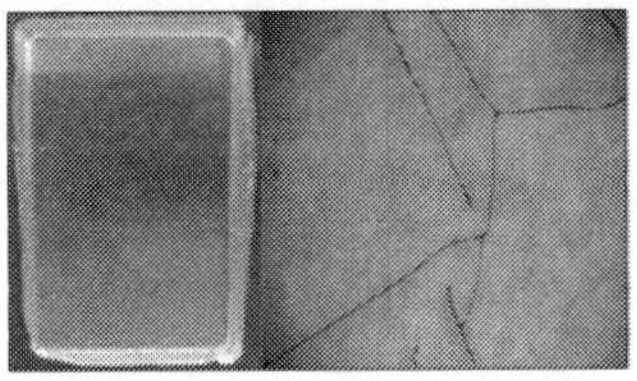

Figure 15. Result of 2600kgf Hot-Embossing

3.7 Analyze the influences of hot-embossing time to defects on the surface

Among parameters setting in the experience, time for preheating is 60 seconds, temperature for preheating is 160℃, pressures for preheating is 700 kg. Besides, time of keeping pressures is 60 seconds, temperature for keeping pressures is 65℃, and pressures for keeping pressures is 2900kgf. Furthermore, time for Hot-Embossing is 900 seconds, and temperature for Hot-Embossing is 160℃. Main testing parameters of Hot-Embossing were separated into 4 groups.

3.8 Analyze the parameters of hot-embossing time

As shown in figure 16, when the Hot-Embossing time set as 600 seconds, the surface average roughness value (Ra) is 0.22μm, average roughness depth (Rz) is 2.06μm, and the gap between the maximum and the minimum value (Rdc) is 5.37μm.

As shown in figure 17, when the Hot-Embossing time set as 900 seconds, the surface average roughness value (Ra) is 0.21μm, average roughness depth (Rz) is 1.37μm, and the gap between maximum and minimum value (Rdc) is 1.80μm.

As shown in figure 18, when the Hot-Embossing time set as 1200 seconds, the surface average roughness value (Ra) is 0.23μm, average roughness depth (Rz) is 1.38μm, and the gap between maximum and minimum value (Rdc) is 1.73μm.

As shown in figure 19, when the Hot-Embossing time set as 1500 seconds, the surface average roughness value (Ra) is 0.20μm, average roughness depth (Rz) is 1.10μm, and

https://doi.org/10.1051/matecconf/201925702006

the gap between maximum and minimum value (Rdc) is 1.50μm.

From the result above, Hot-Embossing time has tiny influence upon the roughness of surface appearances, which means that the surface is still smoothness in a certain extent. The inadequate time for crystallizations to weld completely when the shortness of Hot-Embossing time according to figure 16. On the other hand, too much time for crystallizations welding will make the bad appearances, like the residual glue, to the product. It is because the heating and cooling systems were not designed in up mold, the temperature difference cause the residual glue only happened at the side of up mold.

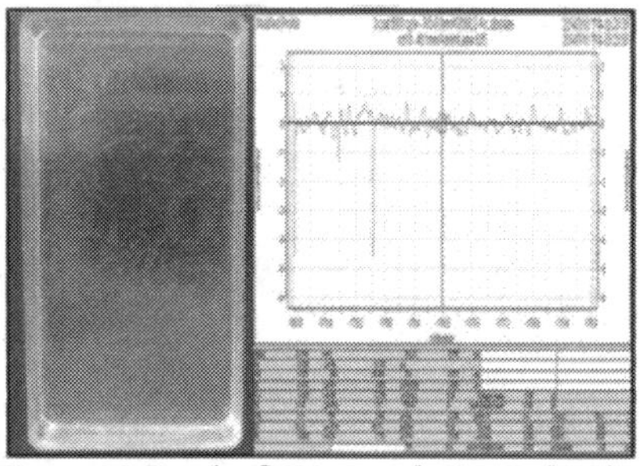

Figure 16. Result of 600 seconds Hot-Embossing

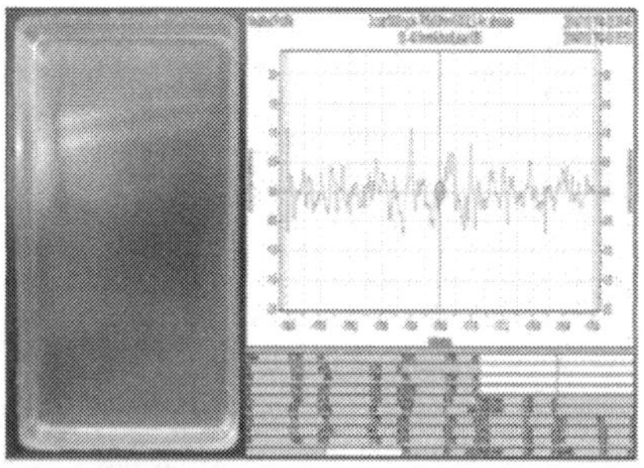

Figure 17. Result of 900 seconds Hot-Embossing

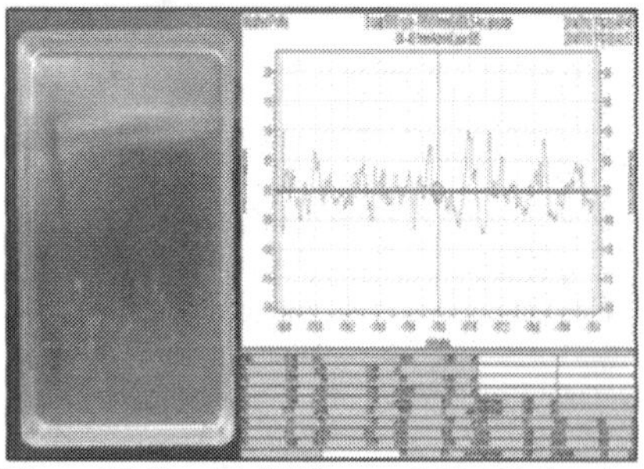

Figure 18. Result of 600 seconds Hot-Embossing

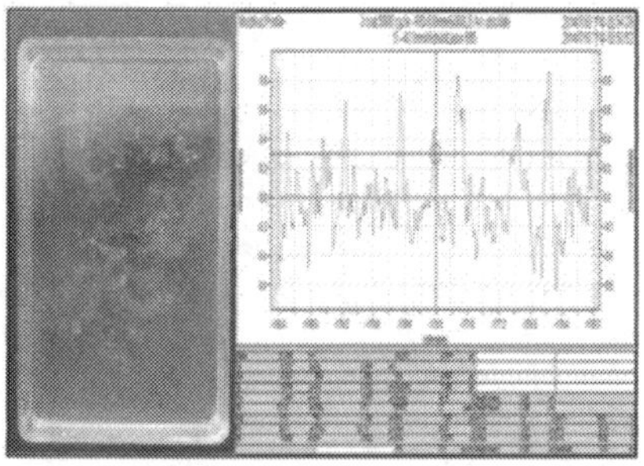

Figure 19. Result of 900 seconds Hot-Embossing

3.9 Analyze the pressure residual stress

According to figure 20, detecting the shell product in the experience through photoelastic instruments would find the low residual stress of product in Hot-Embossing process. There is only 5.37nm in average photoelastic value and 2.68 MPa for average stress value. Thus the warps on product in Hot-Embossing are obscure.

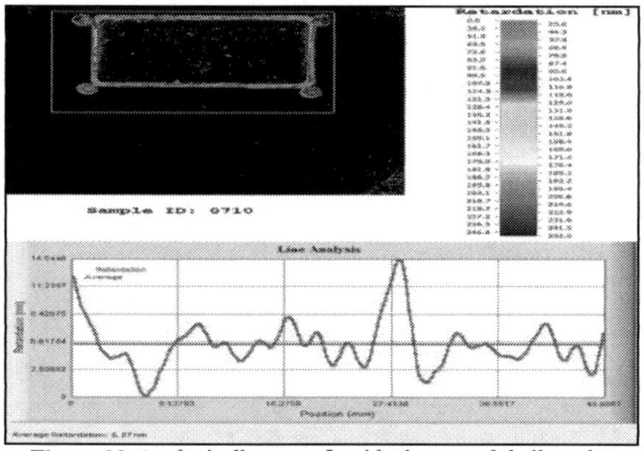

Figure 20. Analysis diagram of residual stress of shell product

4 Discussions and results

4.1 Conclusions for testing positioning technique

Setting 30μm for each adjustment on X axis, the average adjusting process is 29.96μm, and the positioning accuracy is 99.80%. Then setting 30μm for each adjustment on Y axis, the average adjusting process is 29.72μm, and the positioning accuracy is 99.00%. Setting 12', 0.2°, for each rotation adjustment on C axis, the average adjusting process is 0.21°, and the positioning accuracy is 105%. The average error between actual measured value and theory value is only 14.21μm on the reconstruction of product.

4.2 Conclusions for applying positioning technique

The value of actual movement on angle is 22.78' after one-off adjustment; it had a 0.78' error compare with the adjusting value forecast 22'. It is 2' smaller than setting resolution. The value of actual movement on horizontal X axis is 161.29μm after one-off adjustment; it had a 1.29μm error compare with the adjusting value forecast 160μm. It is 10μm smaller than setting resolution. And the value of actual movement on horizontal Y axis is 148.64μm after one-off adjustment; it had a 1.36μm error compare with the adjusting value forecast 160μm. It is 10μm smaller than setting resolution. It can prove that the Hot-Embossing machine with hybrid positioning functions which built in the experience is confirmed to the result of positioning and adjusting technique mentioned above after one-off positioning adjusting capability. It possesses the ability of speedy adjusting and positioning.

5 Conclusion

The study combined the features of Hot-Embossing process, positioning techniques, and wafer packaging together; also referred to the literature review to design and develop the Hot-Embossing machine with hybrid positioning functions. Then, evidences of its abilities are gained in the experiences. Comparing to the dual-side positioning plat form of Chiu[5], the accuracies are much more improved. Positioning accuracies of X axis directions from 77.47% achieved to 99.80%, Y axis directions from 90.47% to 99.00% under the process with relative media which could bring errors and the factors of manual operation. Also, positioning accuracy of C axis direction which utilized principle of vernier caliper achieved 105% from 100.02%. Among them, the combination of C axis rotary table and movable side board reduced the volume and weight significantly. Only the angular adjusting resolutions still has deviations after angle and rotation unit transforming. Also, because the scale table is transferred on the circumference of turntable, it needs the conversion from formula. Therefore, the tiny gaps between adjusting value would accumulate. Besides, according to the product and average error from testing of reconstructions are only 14.21, which the average ration of reproduction is 98.57%, it can actual apply in Hot-Embossing process via characteristic of strong spring parallel.

References and notes

1. Cheng-Hsien Wu, Cheng-Hao Chiu, "Hot Embossing with Two-Axis Micro-Positioning for Double-Sided Microstructures", Advanced Science Letters, Vol.4, pp. 1308-1311 (2011).
2. D. Yao and B. Kim, J. Micromech. Microeng. 12, 604 (2002).
3. M. Heckele, W. Bacher and KD Müller, Microsystem Technologies 4, 122 (1998).
4. Cheng-Hsien Wu and Hsien-Chang Kuo, Journal of Mechanical Science of Technology 21(10), 1338 (2007).
5. Cheng-Hsien Wu, Cheng-Hao Chiu, "Research and Development about Error Analysis of Hot-Embossing for Double-Sided Microstructures", The 2nd International Symposium on Hybrid Materials and Processing(2011) October 26-29; Busan, Korea.

A Novel Process Model of Ship Rust Removal by Premixed Abrasive Jet based on Neural Network

Qing Guo[1], Shuzhen Yang[1,2] and Tao Yu[1]

[1] *School of Mechatronic Engineering and Automation, Shanghai University, Shanghai, China*
[2] *School of Intelligent Manufacturing and Control Engineering, Shanghai Polytechnic University, Shanghai, China*

Abstract. In view of the technological requirements of the development of green shipbuilding technology on the effect of ship surface rust removal, the premixed abrasive jet technology is used to remove rust. Because the rust removal of ships with premixed abrasive jet is influenced by multiple parameters and has a high nonlinear relationship between various parameters, the accurate process model of it is difficult to establish. On the basis of artificial neural network modelling technology, the model of ship rust removal with premixed abrasive jet is built. The model takes the system pressure, the target distance, the moving speed of the spray gun and the particle size of the abrasive as input parameters, and the score which can most reflect the effect of the rust removal as output parameter. The test results show that the prediction error of the model is small, and it can better reflect the process rule between the effect of the premixed abrasive jet and the process parameters. We can guide the selection of process parameters according to the model.

1 Introduction

Due to the special working conditions of high salinity and high humidity, the corrosion degree of the surface of the ship is generally serious and the corrosion area is large. It is very difficult to effectively remove rust on its surface, and not pollute the river where the ship is located at the same time [1]. The premixed abrasive jet technology is widely concerned because of its high rust removal efficiency, no pollution and low equipment requirements [2]. Its abrasive particles and water medium are mixed evenly, and the quality of jet is superior to that of other jets [3]. But so far, the setting of the parameters before processing depends on the experience of the rust removal workers and the best parameters obtained by the single factor experiment [4-5], which ignores the coupling effect between the various process parameters and is not universal.

Artificial neural network is a complex network system which has the characteristics of flexible self-learning, self-organization and self-adapting. Through network learning, it can approximate any linear or nonlinear functions [6]. In this paper, artificial neural network modelling technology is used to build the model of ship rust removal with premixed abrasive jet. The effect of the ship rust removal is predicted based on the established processing parameters, and the model is used to guide the selection of the processing parameters conversely.

2 The principle of ship rust removal by premixed abrasive jet

The principle of the ship rust removal device is shown in Figure 1. In order to ensure the continuous feeding of the ship rust removal system of premixed abrasive jet, the working mode of the double storage tank is used alternately. Before work, the mud pump and control valves are used for feeding. During the work, the water source is added to the abrasive tank. At the same time, it is injected into the mixing chamber too, so that the liquid-solid two phase flow which flows into the mixing chamber from the abrasive tank is mixed fully. Finally, through the acceleration of the pipeline, the abrasive is sent to the spray nozzle on the surface of the ship's steel plate to carry out the ship rust removal operation. With the kinetic energy of high pressure water, the abrasive particles are constantly impacting and grinding the hull on the surface of the hull, and the quality and efficiency of the jet are high. Experiments and the collection of sample data in this paper is based on the equipment which is independently developed. Considering the factors such as hardness, strength, toughness, rust removal effect, rust removal efficiency and cost, the ship rust removal device decides to use river sand that does not cause environmental pollution in the river as abrasive particles.

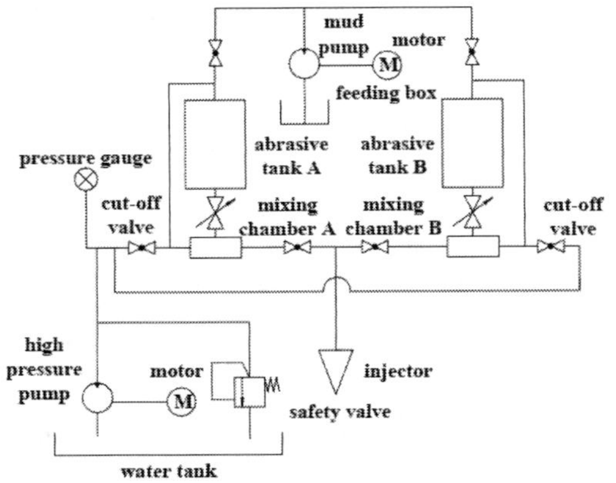

Figure 1. Principle of the ship rust removal device.

3 Process model of rust removal by premixed abrasive jet

3.1 Determination of the parameters of the input layer / output layer

There are many factors affecting the process of rust removal, such as pressure, flow, abrasive, nozzle, injection angle, jet target distance, jet speed and so on. The effect of rust removal is the result of the common effect of these factors. In order to simplify the model, combining the experience of the rust removal equipment operators and related information, finally, it is determined that the input parameters of the model are system pressure, target distance, moving speed of spray gun and abrasive particle size [7].

Sa is used to represent the effect of the jet injection on the rust removal in general, which is divided into 4 grades of Sa1, Sa2, Sa2.5 and Sa3. In order to quantify the effect of rust removal, corresponding the above 4 grades to a different score, as shown in Table 1. The model is established by using the score that can most reflect the effect of rust removal as the output parameter of the model.

Table 1. Evaluation standard of rust removal effect.

Grade	Score	Evaluation criterion
Sa1	1	There is no visible oil, dirt and insecure attachment on the surface of the steel plate.
Sa2	3	There is no visible oil or dirt on the surface of the steel plate. Basically scavenging the oxide skin, rust and other attachments and has strong adherent residue.
Sa2.5	4	There is no visible oil, dirt, and oxide, rust, paint and other attachments on the surface of the steel plate. Only trace or striped spots are retained.
Sa3	5	The surface of the steel plate has no visible oil, dirt, and oxide, rust, paint and other attachments, with a uniform metallic lustre.

3.2 Determination of model structure

There are many types of neural networks. BP neural network is a kind of artificial neural network which uses error back propagation. It is the most widely used. In this paper, based on BP neural network, the process model of ship rust removal with premixed abrasive jet is constructed. The specific formula is as follows:

$$y = g\left(\sum_{j=1}^{m}\left(W_j f\left(\sum_{i=1}^{n} V_{ij} X_i + b_j\right) + b\right)\right) \quad (1)$$

Where x represents input vector, y represents output vector, i represents the number of variables in the input layer, in this article is 4, j represents the number of neurons in the hidden layer, V_{ij} represents the weight matrix between the input layer and the hidden layer, W_j represents the weight matrix between the hidden layer and the output layer, b and b_j represent neural network threshold, f and g represent transfer function and activation function. In this paper, the sigmoid tangent hyperbolic tangent function and the purelin linear transformation function are used as the transfer function and the activation function.

In neural networks, an appropriate increase in the number of hidden layers can improve the learning and generalization ability of the model. But the error signal is transmitted from the hidden layer to the input layer by the output layer. The more hidden layers, the more unreliable when the error passes to the input layer. Referring to related literature [8], we can see that a single hidden layer neural network with enough neurons can approximate any function which is smooth, measurable between input and output by selecting appropriate connection weights and transfer functions. In order to simplify the model, this paper decides to adopt one hidden layer. The structure of BP neural network for ship rust removal with premixed abrasive jet as shown in Figure 2.

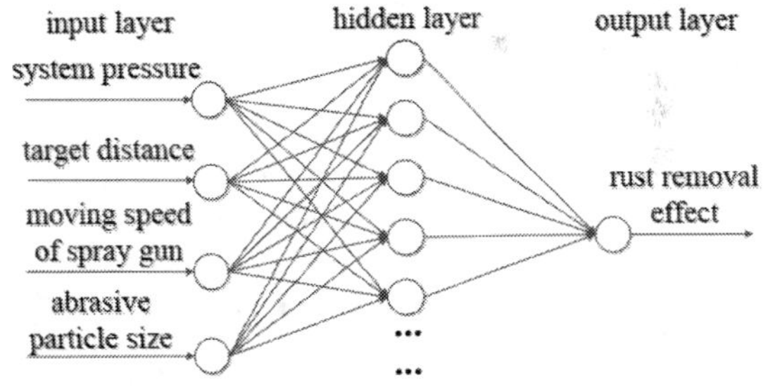

Figure 2. Network structure of BP artificial neural.

3.3 Algorithm and structure optimization

The BP neural network itself has some shortcomings, which is mainly reflected in the slow convergence of the learning process and easy to fall into local minimum. In view of the above defects, many scholars have proposed different improved algorithms. In this paper, the L-M algorithm (Levenberg-Marquardt) is used to improve the stability of the model. L-M algorithm adopts numerical optimization technology, combines the advantages of gradient descent method and Gauss Newton method. It can not only improve the convergence speed of network, but also reduce the possibility of falling into local minimum.

The number of neurons in the hidden layer will directly affect the accuracy and generalization of the model. If the number of neurons is too small, it will lead to lack of fitting, and the network can't describe the data accurately. If the number of neurons is too large, it will lead to excessive fitting and affect its performance. The number of neurons can be determined by comparing the mean relative error of different neuron numbers, it can also be determined by the empirical formula of hidden layer neurons:

$$M = (N + P)^{1/2} + c \quad (2)$$

Where c takes the integers between 1 and 10, N represents the number of neurons in the input layer, in this paper is 4, P represents the number of neurons in the output layer, in this paper is 1, M represents the number of neurons in the hidden layer, takes the integers between 3 and 13.

4 Experiments

4.1 Acquisition and pre-processing of sample data

There are many factors affecting the effect of rust removal. In order to eliminate the influence of other factors on the network, it is necessary to ensure the consistency of the other factors in each experiment. Through preliminary experiments, it is found that the coupling effect between the injection angle and the other factors is smaller. Refer to the relevant literature, and the verification of the single factor experiment, it is found that the ship's rust removal effect is the best when the angle of the jet and the corroded steel plate is 60 degrees [9]. The relevant parameters of the experiment are set as follows:

(1) System pressure. According to the preliminary experiment, the effect of rust removal is better when the pressure of the system is over 7 MPa. However, if the pressure is too large, the gun recoil will increase quickly, the gunman is difficult to grasp which is danger. Therefore, the initial system pressure is between 7 and 11 MPa.

(2) Target distance. The target distance refers to the distance between the jet nozzle and the surface of the ship's steel plate, it is best to be fixed in the area of the energy concentration of the jet. If the target distance is too large, it will lose more power and energy of the jet, and can't play a very good damage to the surface of the ship. The initial target distance is from 300mm to 500mm.

(3) Moving speed of the spray gun. When the moving speed of the spray gun is too small, it can better remove the rust and the attachment, but the efficiency is not high. In order to ensure the rust removal efficiency, the initial moving speed of the spray gun is 300 ~700 mm/min.

(4) Abrasive particle size. The particle size of the abrasive has a certain influence on the quality and efficiency of the rust removal. According to the experience of the operators and preliminary experiments, mesh number is taken for the 16, 18, 20, 24 and 28.

In summary, part of the sample data is shown in Table 2. Because neural network is sensitive to sample data, the order of sample data and the difference between numerical values of different parameters will affect the training time and accuracy. Practice has proved that in the process of network training, we first normalize sample data and import them into neural network for training, which can improve the efficiency, accuracy and generalization ability of network training. The premnmx function is used to normalize the sample data.

4.2 Results and analysis

Based on the MATLAB R2015b software platform, the functions of newff and train in the artificial neural network toolbox are called, the training and testing of the network are carried out. Training with the first 40 groups of data, then testing with the last 10 groups of data. The relationship between the number of neurons in the hidden layer and the average relative error of the prediction results is shown in Table 3. It is known that when the number of neurons is 8, the average relative error is the smallest, which is 1.34%. Therefore, it is decided to adopt the network structure of 4-8-1.

Table 2. Part of the sample data.

	System pressure(MPa)	Target distance (mm)	Moving speed of the spray gun (mm/min)	Abrasive particle size (mesh number)	Rust removal effect
1	7	300	300	16	3.97
2	7	350	400	18	4.01
3	7	400	500	20	4.14
4	7	450	600	24	3.86
5	7	500	700	28	3.71
6	7	300	400	24	3.94
7	7	350	300	28	4.07
8	7	400	600	16	3.76
9	7	450	700	18	3.81
10	7	500	500	20	4.11
11	8	300	500	16	4.12
...					
50	11	500	300	28	4.71

Table 3. The relationship between the number of neurons in the hidden layer and the model error.

The number of neurons in the hidden layer	Average relative error of prediction
3	2.85%
4	1.92%
5	1.65%
6	2.91%
7	1.60%
8	1.34%
9	1.73%
10	2.84%
11	1.87%
12	1.68%
13	3.62%

The experiments were done to verify the effectiveness of the premixed abrasive jet ship rust removal process model. The processing parameters and the effect of rust removal were inputted into the model. The prediction results of the model were compared with the actual effect of rust removal. Figure 3 is the comparison curve of the effect of the actual rust removal and the prediction results of the model. The maximum error is 2.41%, the minimum error is 0.03%, and the average relative error is 1.32%.

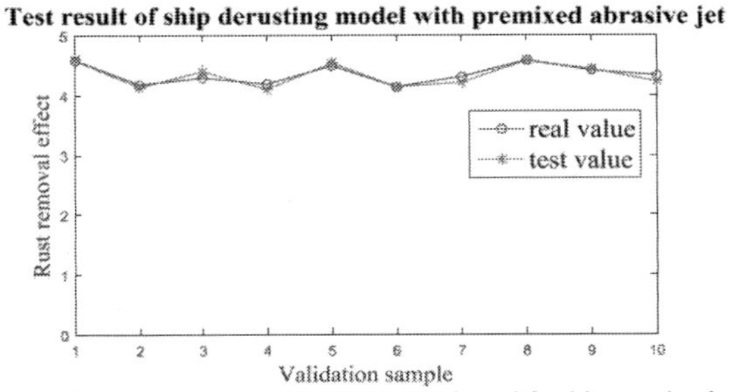

Figure 3. Test result of ship rust removal model with premixed abrasive jet.

Based on this model, taking a steel plate that is equivalent to the rust of the ship to test. After 243s, the corrosion of the steel plate was completely removed. The effect of rust removal reaches Sa2.5, and the efficiency of rust removal reaches 14m2/h. Compared with the efficiency of rust removal not based on this model, it has increased by 16.7%.

5 Conclusion

In view of the actual situation of the setting of processing parameters depends on the experience, the ship rust removal process model of premixed abrasive jet based on neural network is established. The prediction model can quickly predict the effect of rust removal when the system pressure, the target distance, the moving speed of the spray gun and the particle size of the abrasive are determined. The prediction model takes "4-8-1" as the optimal structure, with the maximum prediction error of 2.41%, minimum prediction error of 0.06% and the average relative error of 1.34%. In general, it can truly reflect the process rule between the effect of ship rust removal by premixed abrasive jet and the process parameters, we can guide the selection of process parameters based on this model.

Acknowledgments

This work was supported by Shanghai Committee of Science and Technology, China (Grant No.17511107302) and the Project of Key Discipline of Shanghai Polytechnic University (No. XXKZD1603).

References

1. Xie Yufei, Liu Hongwei, Hu Yongxiang. Study on the method of determining the technological parameters of laser rust removal of ship plate [J]. *Chinese laser*, **4**, 103-110 (2016).
2. Liu Guoyong, Chen Xinxin, Zhu Dongmei, et al. Internal flow field of mixed abrasive jet mixing cavities of premixed abrasive jet [J]. *Journal of University of Science and Technology Beijing*, **4**, 830-837 (2014).
3. Zuo Weiqin, Lu Yiyu, Liu Mingju, and so on. A new method of testing the abrasive speed of the premixed abrasive jet [J]. *Journal of Chongqing University: Natural Science Edition*, **39**, 96-102 (2016).
4. Zhou Mengying. Application of premixed abrasive jet technology to pretreatment of anticorrosion of large metal components in coal mine [J]. *Coal engineering*, **1**, 94-96 (2008).
5. Yang Gang. Simulation and study of abrasive water jet rust removal technology [J]. *Mechanical and electrical engineering*, **30**, 929-932 (2013).
6. Liu H, Wang J, Kelson N, et al. A study of abrasive waterjet characteristics by CFD simulation [J]. *Journal of Materials Processing Technology*, **1**, 488-493 (2004).
7. Liu L H, Cao H, Zou C Q. System of Premixed Abrasive Jet Cleaning and Derusting on Steel Tube In wall [J]. *Applied Mechanics & Materials*, **4**, 2571-2574 (2012).
8. Shi Lingzhi, Deng Qihong, Lu Chan, et al. BP artificial neural network based on PM_ atmospheric particles (10) concentration prediction [J]. *Journal of Central South University (NATURAL SCIENCE EDITION)*, **43**, 1969-1974 (2012).
9. Nguyen T, Shanmugam D K, Wang J. Effect of liquid properties on the stability of an abrasive waterjet[J]. *International Journal of Machine Tools & Manufacture*, **48**, 1138-1147 (2008).

MATEC Web of Conferences **257**, 02008 (2019) https://doi.org/10.1051/matecconf/201925702008
PMAE 2018

Tip-replaceable Horn with Full Wavelength for Ultrasonic Welding

Mi Ae Kim , Chang Seo Im and Dong Sam Park

Dept. of Mechanical Engineering, Incheon National Univ., Yeonsu Gu, Incheon, 22012, Korea

Abstract. Ultrasonic welding is faster, environmental friendly, and more efficient than conventional welding. Ultrasonic welding stack, the key part in ultrasonic welding equipment, is composed of transducer, booster, and horn. In the case of ultrasonic welding, wear in the welding tips or breakage of horn occurs because ultrasonic vibration of tips is transmitted directly to the welding material. If grinding of the tips is repeated several times, the mass of horn is reduced and the frequency characteristic is changed. So, there is a limit to refurbishment. In additions, as materials of horn are relatively expensive, some cost problems could be occurred for replacing horns. This study, in order to reduce the manufacturing cost of horn, devised a tip-replaceable ultrasonic horn so that its tip can be easily replaced whenever it is worn. For the designed shape, natural frequency and vibration mode are analyzed using FEA. Through the FEA, the optimal tip-replaceable horn with resonance frequency of 40 kHz is proposed.

1 Introduction

Ultrasonic welding is a faster, more environmentally friendly and simpler, more efficient welding method than conventional welding. In particular, ultrasonic metal welding has the advantage of being able to bond dissimilar metals such as aluminum and copper [1]. In addition, it is an essential technology for miniaturization and weight reduction of electrical products and automobile manufacturing [2] and is known as the most suitable method for electrode terminal welding of rechargeable batteries used in automobiles and mobile phones.

When two metals are brought into contact and one metal is vibrated in parallel, they are bonded in a short time [3]. Only the vibrational energy is used and no current or external heating is required [4]. Welding occurs at relatively low temperatures (<300 ° C) [5], and high-frequency shear vibrations applied under appropriate pressure also have the advantage of removing oxide and contaminant layers [6].

Ultrasonic welding stack, a key component of ultrasonic welding equipment, consists of transducer, booster and horn. The tips knurled in the horn main body require direct contact with the welding material and transmit ultrasound vibrations, which causes tip wear during welding. In this case, re-grinding decreases the mass of the horn and changes the frequency characteristics. Therefore, there is a limit to re-grinding. Also, horns are made of expensive materials such as titanium alloys, which can be expensive if we are to replace horns.

To reduce the cost of the horn, a patent for an ultrasonic horn that can be mounted on the body of the horn in the form of a tip plate has been submitted in Korea [7]. Nissan Motors in Japan also applied for a patent regarding an ultrasonic welding horn that allows the tip to be attached and detached by screws between them. Branson, USA, applied for a patent for an ultrasonic welding horn that can replace a tip using a collet. However, most of these patents relate to half-wave length horns. Although a one-wave replaceable horn has been patented in Japan [8], but its structure is designed to replace almost half of the horn. To overcome this problem, a new horn design is required to minimize the mass of the replacement part and increase efficiency. However, despite the importance of tip-replaceable horns, there is not much academic study of them till now. In this study, we designed a tip-replaceable ultrasonic welding horn with resonance frequency of 40 kHz and analyzed vibration characteristics using FEA (finite element analysis).

2 Designing a tip-replaceable horn

2.1 Initial model of tip-replaceable horn

Fig. 1(a) is a 3D model of conventional one-body horn for ultrasonic welding, and Fig. 1(b) is an initial design of the tip-replaceable horn proposed in this study. One of the tip parts is dug out from the conventional metal welding horn, and a rectangular shaped tip is bolted to the horn body. The width of the space for the assembled tip is 18.9 mm, equal to the width of the assembled tip, and its depth is 3.9mm, equal to the thickness of the tip. So, the left and right sides of the replaceable part are in contact with the horn, and the upper surface of the assembled tip and surface of the horn are both in the same plane.

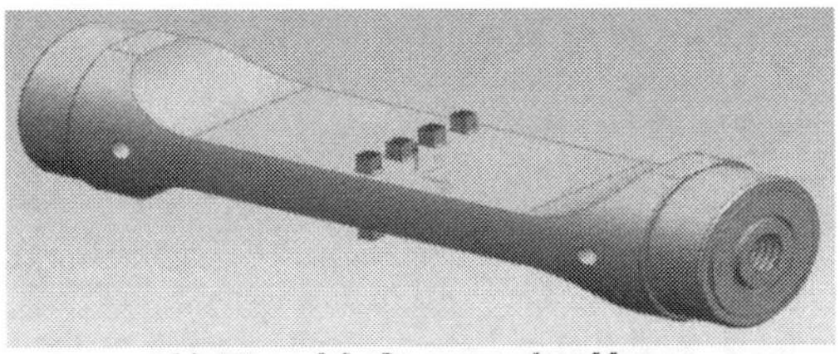

(a) 3D model of a conventional horn

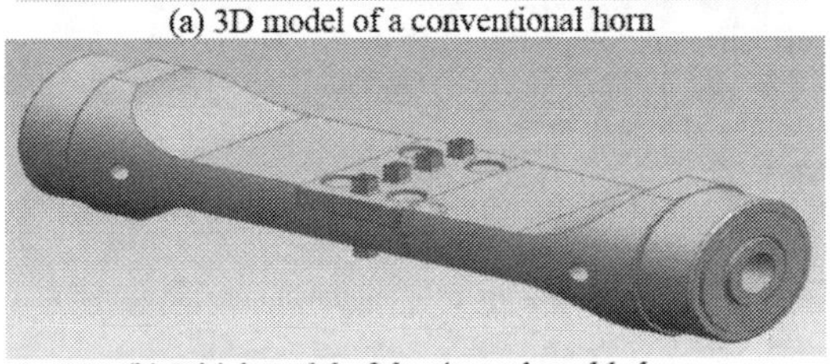

(b) Initial model of the tip-replaceable horn

Figure 1. Models of conventional horn and the tip-replaceable horn proposed in this study

2.2 Analysis of vibration characteristics

Finite element analysis was carried out using the tip-replaceable horn model shown in Fig. 1(b). Modal analysis was conducted to analyze natural frequency and vibration mode using the ANSYS tool. Horn material was assumed to be ASP60, and its material properties are ; density 7.9 g/cm^3, Poisson's ratio 0.3, and Young's modulus 222 Gpa.

As contact conditions, "Bonded" was applied to parts that fixed with threaded bolts, and "No Separation" was applied to the parts that were not completely fixed. As support conditions, "Cylindrical support" was applied to bolt hole and the degree of freedom was given only in the axial direction. Vibration modes from 1st to 20th were analyzed. The 14th mode showing the first longitudinal vibration mode is shown in Fig. 2. Its natural frequency is 38,067 Hz, which is about 0.5 kHz less than the natural frequency (38,526 Hz) of the one-body horn.

Figure 2. 14th vibration mode (natural frequency 38,067 Hz)

3 Improving horn design by modal analysis

3.1 Vibration mode of the initial model

As there are no contact conditions between the right and left sides of the tip and the horn body, the analysis was carried out without any condition in those parts. In consequence, Interference occurred between tip and horn body as shown in Fig. 2. This indicates that the vibrations of the tip and horn are not exactly the same, and interference may occur between the tip and the horn during vibration. In this case, a continuous impact is applied to both the tip and the horn, which can affect the life of the horn and the quality of the weld.

As shown in Fig. 3(a), the initial model is modified with gap 0.5 mm between the tip and the horn. Although the vibration mode 14th of the modified model is similar to the longitudinal vibration mode, as shown in Fig. 3(b), vertical bending vibration occurs slightly. In other words, when the transducer or the booster is assembled on the end surfaces of the horn, its vibration may not be properly transmitted because both surfaces do not contact perfectly. It is considered that slight bending vibration is caused by the problem that the upper and lower shapes of the horn are not completely symmetrical. Fig. 3(c) shows the distribution of the vibration amplitude at the cross section along the tips. The amplitude in the replaceable tips is found to be smaller.

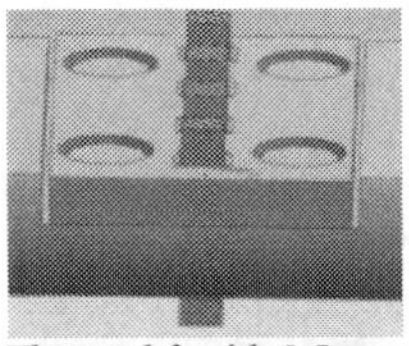

(a) The model with 0.5mm gap

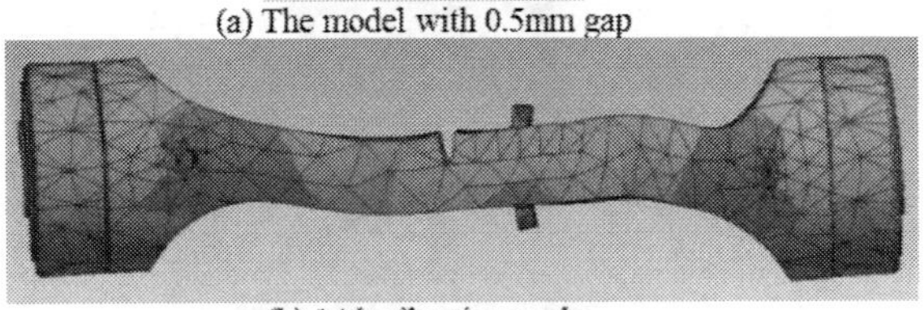

(b) 14th vibration mode

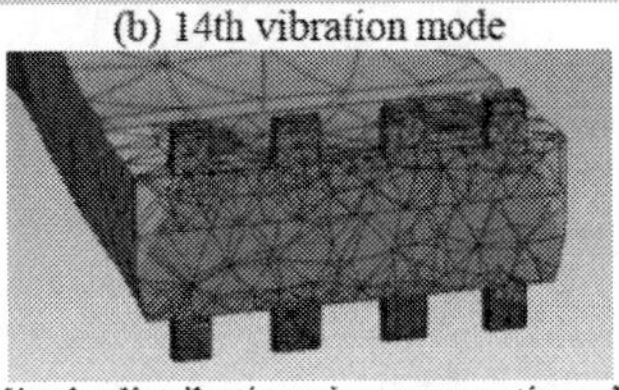

(c) Amplitude distributions in cross section along tips

Figure 3. Modal analysis of the modified horn with 0.5mm gaps

3.2 Improving horn design

As reduction of the amplitude in the assembled tips and uneven amplitude at the side surfaces of the horn are due to the asymmetry shape of the upper and lower sides of the horn, the horn is modified as shown in Fig. 4. Fig. 4(a) shows the horn that top and bottom tips are replaceable. Fig. 4(b) shows the horn that only the upper tip is replaceable, and in the lower part slots are machined to keep the horn as symmetric as possible.

Fig. 5(a) and Fig. 5(b) show vibration modes of Fig. 4(a) and Fig. 4(b), respectably. Both horns show good longitudinal vibration mode, and the amplitude of replaceable tips is relatively large. The vibration amplitude of end-sides of the horn is relatively uniform. In this study, the horn with one replaceable tips is adopted as a final model considering the rigidity of the horn and the convenience of machining.

As shown in Fig. 5(b), however, the natural frequency of the 14th mode was 37,281 Hz, which is much lower than that of the conventional one-body type horn, 38,526 Hz. In order to increase the natural frequency, each thickness of the lower and upper sides of the horn was increased by 1 mm. The total depth of slot in lower side is 3.5mm. Fig. 6(a) shows the horn model modified again, in which the thickness of the area marked in orange is increased. Fig. 6(b) shows the 14th mode shape with a natural frequency of 38,331 Hz. It is confirmed that the mode shape of the modified horn is nearly symmetrical due to the reduction of the protrusion of the tip, so that the vibration occurs almost in the axial direction. The natural frequency of the 14th mode is increased similarly to that of the conventional one-body horn by increasing the mass of the horn.

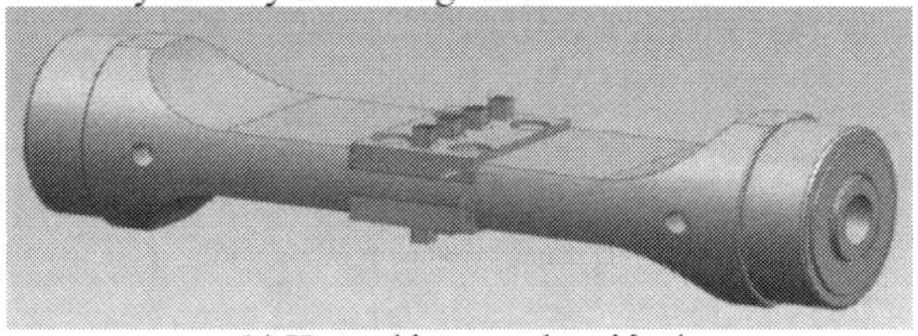

(a) Horn with two replaceable tips

(b) Horn with one replaceable tips

Figure 4. New designs for improving the symmetry

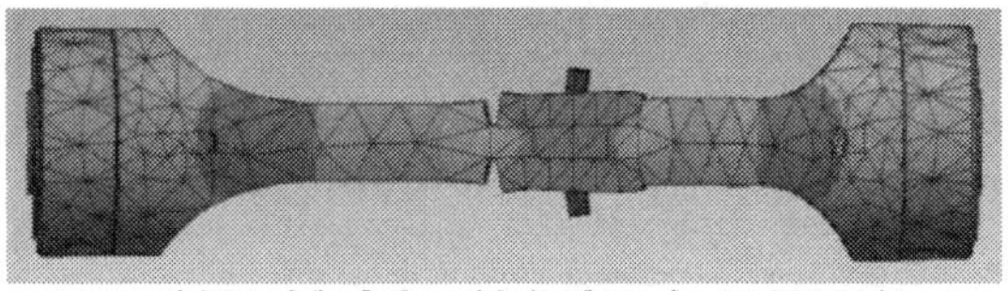

(a) Model of Fig. 4(a) (14th mode, 36,789 Hz)

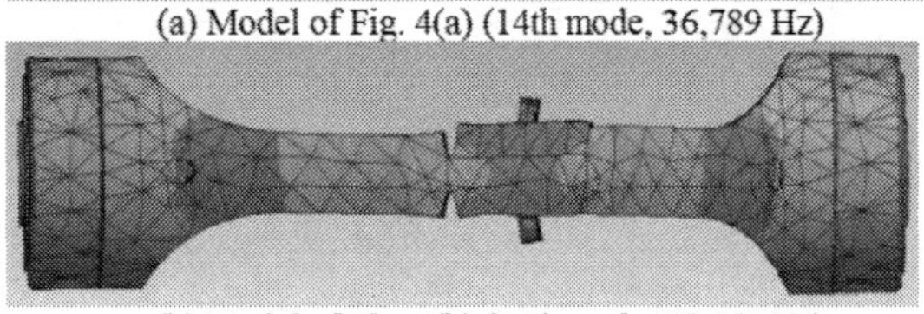

(b) Model of Fig. 4(b) (14th mode, 37,281 Hz)

Figure 5. Vibration modes of new designs

4 Final design and vibration analysis

4.1 Final design

In the process of finding the optimal design of one tip-replaceable horn, it is confirmed that change of the gap width has almost no effects on the vibrational mode shapes. So the gap between tip and horn is designed to be 2mm wide to prevent interference between them. Machined corners are rounded to avoid stress concentration. In the final design, two holes are formed to facilitate pushpin insertion. Two pushpins are used for tight fastening of replaceable tips and horn body in a specified position. Pins are assumed to be assembled without any counterbores on the drilled holes, because there are no significant differences in the natural frequency and mode shape.

The 14th vibration mode and natural frequency of the final model proposed in this study are shown in Fig. 7(a). It is confirmed that vibrations occur almost in the axial direction and the natural frequency is similar to that of the conventional one-body horn. Fig. 7(b) shows distributions of the vibration amplitude at the cross section along joining bolts. It is confirmed that vibrational amplitudes are maximum at the tips and are nearly symmetrical in the upper and lower sides. These results imply that the final design for the tip-replaceable horn could be applied to ultrasonic welding effectively.

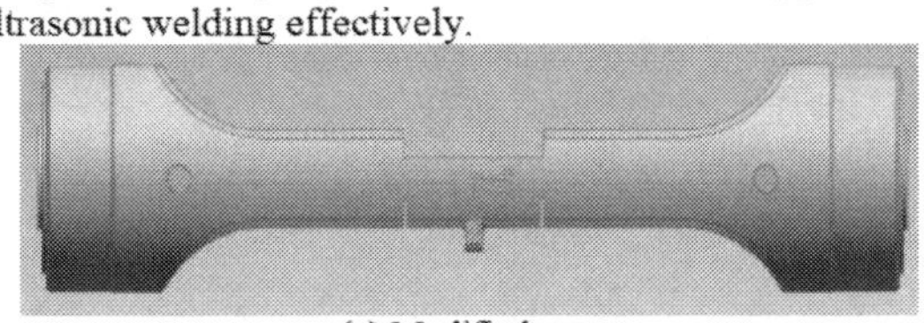

(a) Modified area

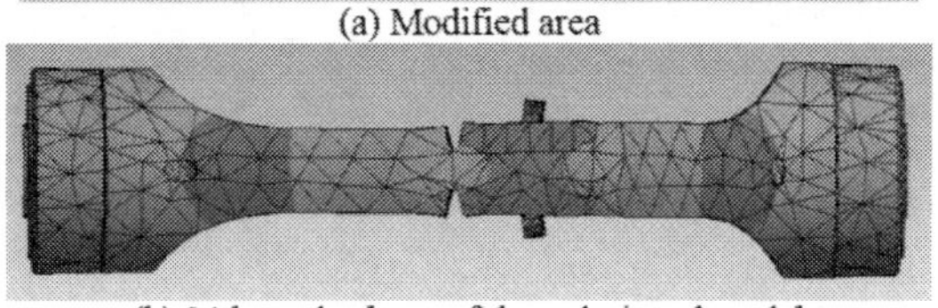

(b) 14th mode shape of the redesigned model

Figure 6. Redesigned model and the 14th mode shape

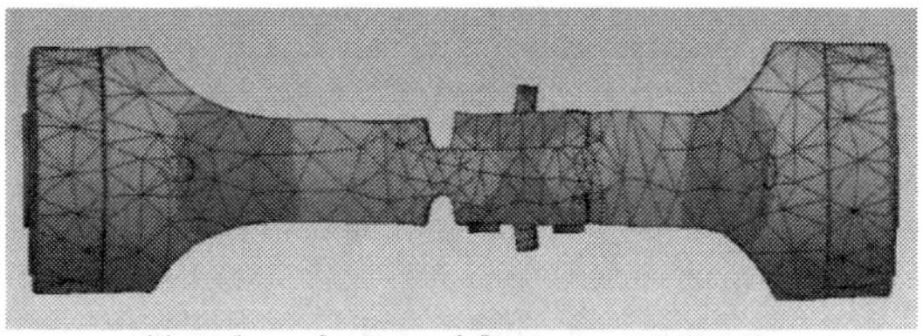

(a) 14th mode, natural frequency 38,326 Hz

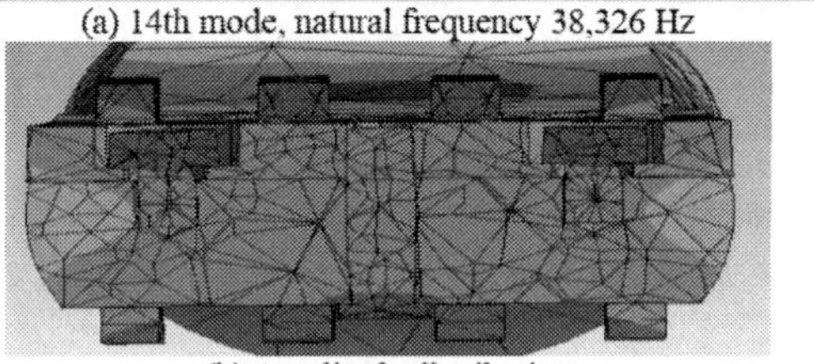

(b) Amplitude distribution

Figure 7. Final FE model and the 14th mode shape

5 Conclusions

The welding tips knurled on a horn body directly contact with welding materials and transmit ultrasonic mechanical vibrations, which necessarily cause wears on the tips during ultrasonic welding. If regrinding of worn tips is repeated several times, the mass of the horn is reduced and the frequency characteristic is changed. So, there is a limit to refurbishment. In this study, the tip-replaceable horn of one-wave length in ultrasound is designed by using FE analysis. This horn can allow the part of the tips to be attached and detached to/from the horn body. Several models of tip-replaceable horn are proposed. The natural frequencies and vibration modes are investigated through FEA.

As the final model, a horn that only the upper tips are replaceable is proposed, and in the lower part of it, slots are machined to keep the horn as symmetric as possible. This model is modified several times to overcome some problems due to the asymmetry shape of the upper and lower sides of the horn. It is confirmed finally that vibrations occur almost in the axial direction and the natural frequency is similar to that of the conventional one-body horn. From analysis of distributions of vibration amplitude at the cross section along joining bolts, it is shown that vibrational amplitudes are maximum at the tips and are nearly symmetrical in the upper and lower sides. These results imply that the final design for tip-replaceable horn could be applied to ultrasonic welding effectively.

Acknowledgement

This work (Grants No. C05118390100470156) was supported by Business for R&D funded The Korea Ministry of SMEs and Startups in 2017.

References

1. Jiromaru Tsujino, "Ultrasonic welding using complex vibration and high frequency vibration," *2011 IEEE International Ultrasonics Symposium Proceedings* (2011), pp. 1910- 1919.
2. Shin-ichi Matsuoka and Hisashi Imai, "Direct welding of different metals used ultrasonic vibration," *Journal of Materials Processing Technology 209*, pp. 954-960.
3. H. S. Jeong, "Fundamentals of Ultrasonic Welding," *Journal of KWS*, Vol. 15, No. 6 (1997), pp. 24-31.
4. E. A. NEPPIRAS, "ULTRASONIC WELDING OF METALS," *ULTRASONICS* (1965), pp. 128~135 Y.C. Chen, D. Bakavos, A.Gholinia, and P.B. Prangnell, "HAZ development and accelerated post-weld natural ageing in ultrasonic spot welding aluminum 6111-T4 automotive sheet," *Acta Materialia 60* (2012), pp. 2816-2828.
5. M. Shakil, N.H. Tariq, M. Ahmad, M.A. Choudhary, J.I. Akhter, and S.S. Babu, "Effect of ultrasonic welding parameters on microstructure and mechanical properties of dissimilar joints," *Materials and Design 55* (2014), pp. 263-273.
6. Sang-gon Kim, "Ultrasonic welding horn has several tips," Patent application number 10-2007-0027299.
7. Nissan Motor Co., LTD, "Ultrasonic welding device," International application number PCT/JP2015/073321.
8. BRANSON ULTRASONICS CORPORATION, "Ultrasonic welding device," Japanese patent application number 2013-89327.

MATEC Web of Conferences **257**, 02009 (2019) https://doi.org/10.1051/matecconf/201925702009
PMAE 2018

Designing of Ultrasonic Horns to Improve Amplitude Uniformity in Ultrasonic Metal Welding

Yeon Ju Lee[1], Muhammad Bilal Shahid[1] and Dong Sam Park[1]

[1]*Dept. of Mechanical Engineering, Incheon National Univ., Yeonsu Gu, Incheon, 22012, Korea*

Abstract. The ultrasonic metal welding has been used extensively in many fields because it can weld thin layers of highly conductive and dissimilar materials with high reliability. In particular, one-wave length horns with four welding tips on both sides of the horn are widely used for welding positive and negative electrodes and tabs of rechargeable lithium ion batteries. As the vibrational amplitude of each tip is known to be non-uniform, this causes some problems in welding performance. In this study, the one-wave length horn which has four welding tips on both sides of the horn is newly designed to improve amplitude uniformity of each tip. The design of experiment technique is employed to determine the optimum configuration of the horn. The modal analysis and harmonic response analysis are repeatedly conducted to obtain the uniform amplitude of horn tips, and stress distribution in the horn is analyzed finally. It is shown that the uniformity of amplitude can be improved by making appropriate grooves on both sides of the conventional horn.

1 Introduction

With the recent expansion of mobile devices, such as smart phones, tablets and PCs, and the development of electric cars, many researches are being carried out actively on their key component, the secondary cell battery. These batteries are supposed to be small, lightweight, and high-capacity. Furthermore, the structure of the battery must be multi-layered. Ultrasonic metal welding has been used for efficient welding of multi-layer tabs.

Most of the studies that have been published are for ultrasonic welding between homogeneous or dissimilar metals [1]. For welding of multiple foil-layers onto a single tab, there is not much research available. So, this topic should be studied in detail. The one-wave length horn which has four welding tips has been widely used to weld positive/negative electrodes and tabs of the rechargeable lithium-ion batteries [2]. The vibrational amplitude of each tip is known to be non-uniform, and this causes some problems in welding performance. Therefore, the differences in amplitude between the tips should be minimized for efficient ultrasonic metal welding.

In past studies associated with this topic, there is a study that reduces the degree of uniformity error that occurs with molded areas during vibratory plate ultrasound of speakers. The vibration characteristics were analyzed by changing the design to improve the uniformity of the output[3]. There is another study that analyzes the effects of grooves and slots on the amplitude uniformity and determines the dimensions of the grooves and slots for high uniformity[4].

In this study, the one-wave length horn which has four welding tips is newly designed to improve vibrational uniformity on each tip of the horn. The design of experiment technique is employed to determine the optimum configuration of the horn. The modal and harmonic response analysis are performed to obtain the uniform amplitude of designed horn tips, and the stress distribution in the horn is analyzed finally. Several different modifications in the shape of conventional horn are tried, and then analyzed by using ANSYS tool.

2 Shape design of horns

2.1 Conventional ultrasonic horns

This study aims to improve weld quality by reducing differences in amplitude between two central tips and two outer tips, which are generated during welding process with conventional ultrasonic welding horns. Fig. 1 shows the ultrasonic welding horn that is being used in industry.

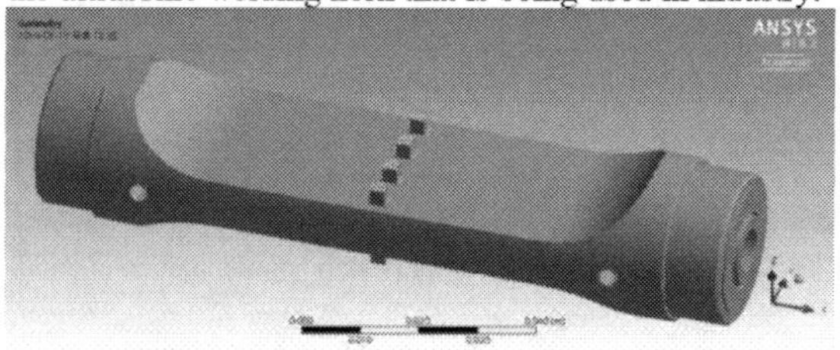

Figure 1. 3D model of a conventional horn

The shape of the horn on which this study is based is a one-wave length horn, designed symmetrically to provide stable welding [5]. The required working frequency of the ultrasonic device is 40 kHz (± 500 Hz) and the device will not operate if the frequency fluctuates beyond the suggested tolerance. Therefore, the natural frequency must be within this range while designing horn. The material of the horn is CPM10 V, which has high hardness and excellent vibration characteristics. Table 1 shows the material properties [6].

Table 1. Mechanical properties of horn material, CPM10V

Density	Poisson's Ratio	Young's Modulus
0.268 [lb/in^3]	0.28	221 [GPa]

2.2 Shape modifying based on uniformity analysis

In the analysis result of the conventional horn, there are differences in amplitude between two outer tips and two center tips, which cause differences in the weld strength at the point of each tip. It makes adverse effect to weldability so the shape of the horn should be modified to prevent it.

Initially, as shown in Fig. 2, A(14mm), B(2.4mm) C(R15) and D(R3) has been chosen as starting values for analysis. With each of the values of A, B, C, and D being varied, only one at a time, the uniformity is measured and compared with the initial model. To check the uniformity, as shown in Fig. 3, the displacements of three points per tip are measured.

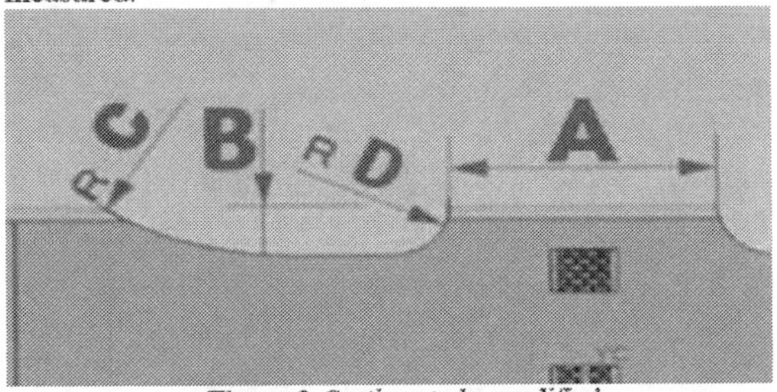

Figure 2. Sections to be modified

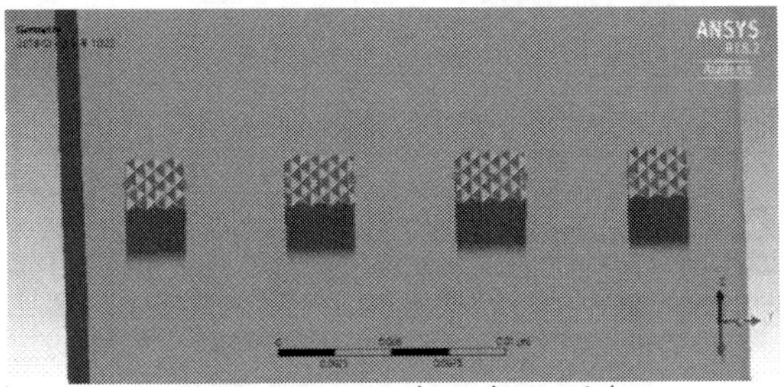

Figure 3. 12 measuring points on 4 tips

The modal analysis was done by increasing the length of part A from 12mm to 16mm. Fig. 4(a) is a graph showing the displacement of each tip. As the most uniform point is found at 14 mm (39,638 Hz) shown by the green line on graph. This length is selected as our design point. The same analysis was done by increasing the length of part B from 2.2mm to 3mm. Fig. 4(b) is a graph showing the displacement of each tip corresponding to different lengths of part B. By changing the radius values from 2.2 to 3.0 mm, the most suitable radius value came out to be 3.0 mm. However, the length of B could not be increased beyond 3mm, because the stress concentration would increase too much.

The same analysis has been carried out for different radius of C and D. As shown as Fig. 5(a), the most suitable radius for C was found out to be R18 mm (39,683 Hz) with the most uniform displacement (purple line). In the graph on Fig. 5(b), R3 (39,683 Hz) was selected as the desired radius owing to the best uniformity at that radius (green line).

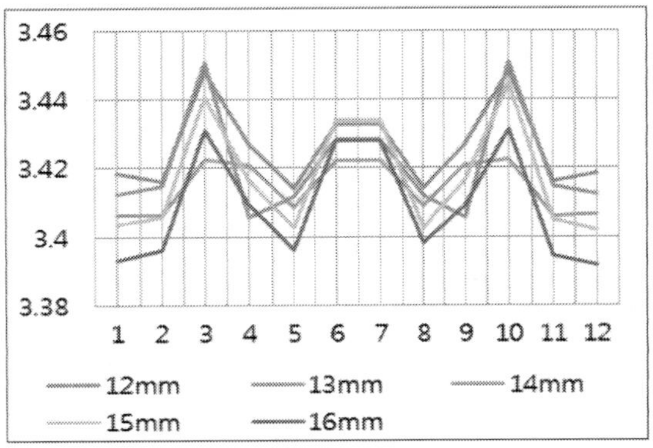

(a) A-length

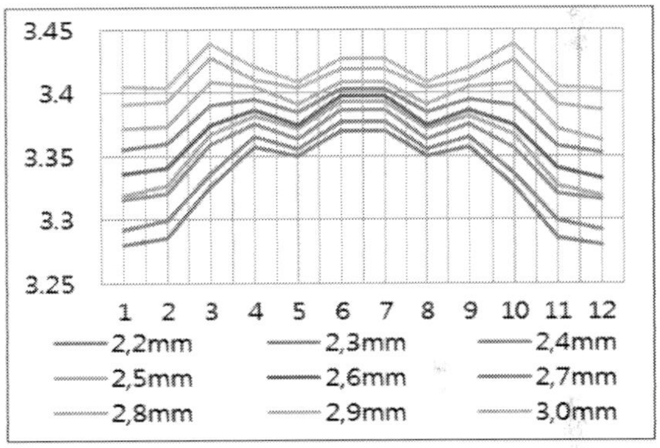

(b) B-length

Figure 4. Effects of the length of A and B on the amplitude variations on the tips

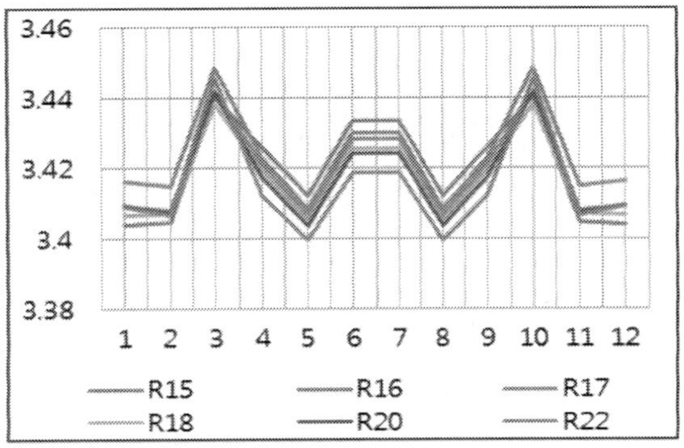

(a) C-radius

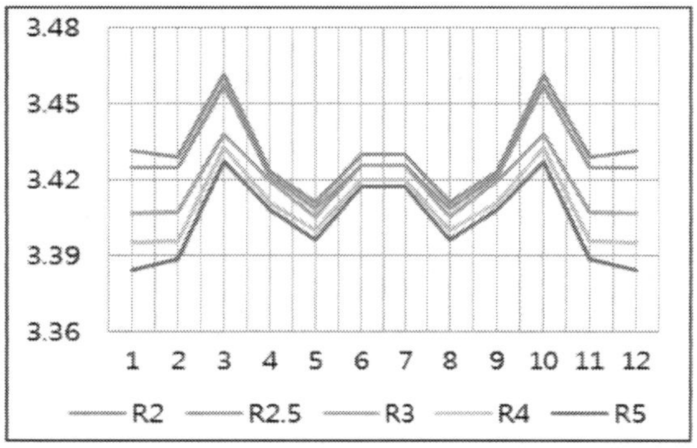

(b) D-radius

Figure 5. Effects of the radius of C and D on the amplitude variations on the tips

3 Vibration and stress analysis

3.1 Modal analysis

To design the new horn, modal analysis is performed using ANSYS tool. Fig. 6 shows non-longitudinal mode (13th, 15th) adjacent to the longitudinal mode (14th). These non-longitudinal vibration modes will adversely affect ultrasonic welding, so it is not suitable for ultrasonic welding that requires the longitudinal vibration mode [7]. The measured frequency in the 14th mode was found to be 39,633 Hz with an error of approximately 0.92 %. The vibration mode is also longitudinal and can be seen to have the maximum amplitude on the center part of horn.

Fig. 7 shows the amplitudes on the tips of horn before (Fig.7 (a)) and after improvement (Fig.7 (b)). As can be seen clearly, higher amplitude (approx. 16.9 ㎛) occurs on tips of the horn on the right one of Fig. 7 i.e. after improvement. Also, the amplitude distribution in the central tips and the outer tips is more uniform after improvement.

(a) 13th mode (37,390Hz)

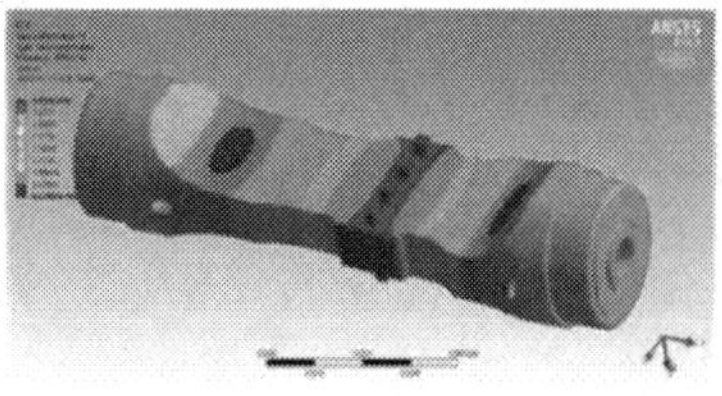

(b) 14th mode (39,633Hz)

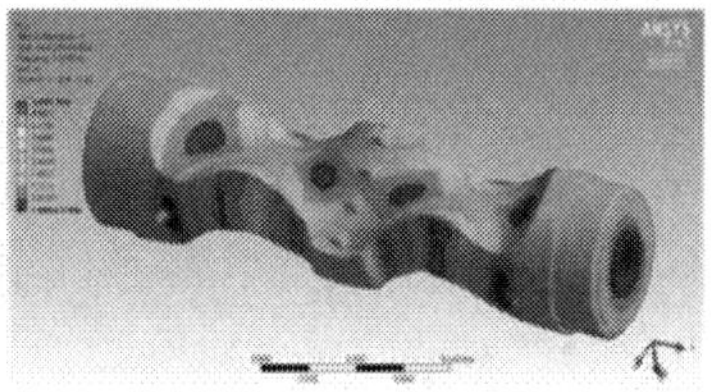

(c) 15th mode (45,290Hz)

Figure 6. Vibration modes and natural frequencies of the improved horn

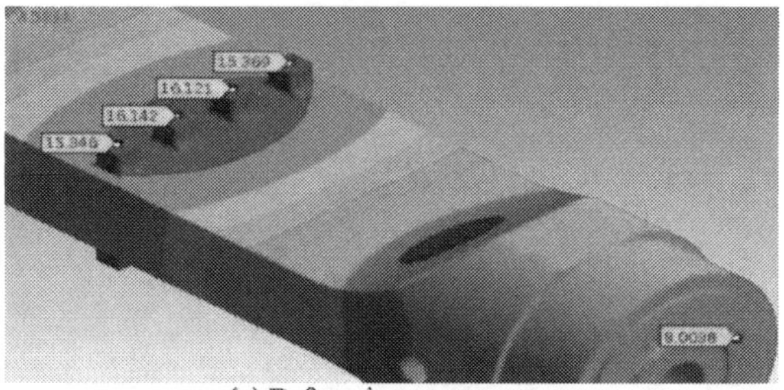

(a) Before improvement

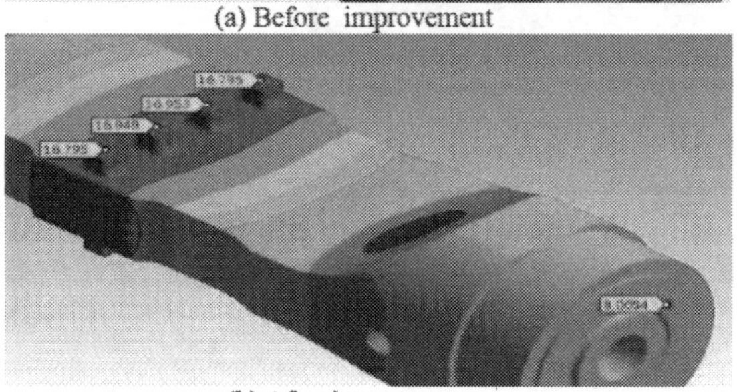

(b) After improvement

Figure 7. Amplitude of horn before/after improvement

3.2 Harmonic response analysis

A frequency range of at least 30,000 Hz to 50,000 Hz is specified for the input surfaces that meet the booster of the ultrasonic welding tool. As a result of the harmonic response analysis to the knurling part of the tip on the output surface, the peak frequency is found to be 39.6 kHz as shown in Fig.8, which corroborates the results of the modal analysis [8].

3.3 Stress analysis

There was no significant difference between the frequencies of the two models before and after the improvements (39,710 Hz). However, as shown in Fig. 9, the maximum stress in the horn was reduced to some extent, which would improve the life of the horn. It was found that the maximum stress was 215.24MPa before the improvement, and after the improvement, the stress level was reduced to 185.4MPa.

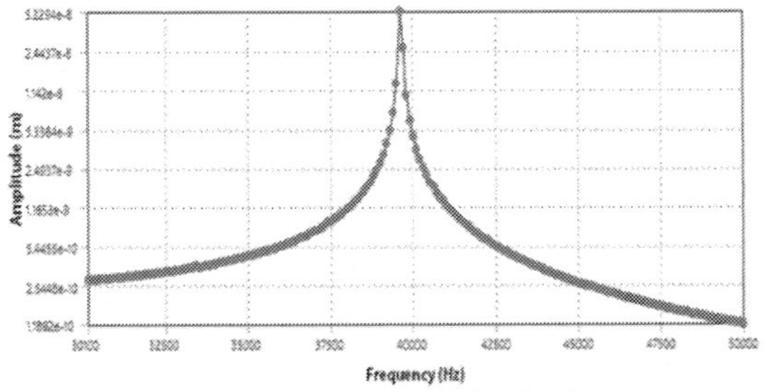

Figure 8. Harmonic response analysis after improvement

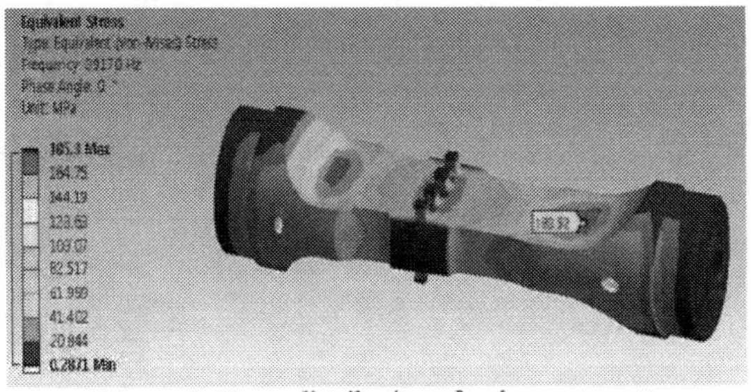

Figure 9. Stress distribution after improvement

4 Summary

Recently, ultrasonic horns of one-wave length which have four welding tips have been widely used to weld positive/negative electrodes and tabs of the rechargeable lithium-ion batteries. The vibrational amplitude of each tip is known to be non-uniform, and this causes the some problems in welding performance. In this study, conventional horn of the one-wave length which has four welding tips is modified and designed newly to improve vibrational uniformity of each tip of the horn. The design of experiment technique is employed to determine the optimum configuration of the horn. The modal analysis and harmonic response analysis are conducted to obtain the uniform amplitude of designed horn tips, and stress distribution in the horn is analyzed finally. It is concluded that the uniformity of vibrational amplitude can be improved by making proper grooves on the both sides of the conventional horn. There was no significant difference between the frequencies of the two models before and after the improvements. However, the maximum stress in the horn was reduced to some extents, which could improve the life of the horn. It was found that the maximum stress was 215.24MPa before the improvement, and after the improvement, the stress level was reduced to 185.4MPa.

Acknowledgement

This research was supported by Basic Science Research Program through the National Research Foundation of Korea (NRF) funded by the Ministry of Education (Grant No. 2017R1D1A1B03034483).

References

1. Ho-Su Jang, Woo Yeol Park and Dong-Sam Park, "Design and weld ability verification of the 40 kHz horn for ultrasonic metal welding," *Journal of the Korean Society of Manufacturing Process Engineers*, vol. 12 (2013), pp. 55-60.
2. Jeong-Seok Seo, Sung-Min Jang and Si-Young Beck, " One-wavelength ultrasonic horn design for ultrasonic machining of mobile phone battery terminal welding," *Journal of the Korean Society of Manufacturing Technology Engineers*, Vol. 21(2012), pp. 70-75.
3. H.J.Lee, Y.E.Lim and K.Park, "Optimal design of ultrasonic horn for thermoforming process," *Journal of Korean Society of Precision Engineering* (2016), pp. 234-234.
4. S.R.Kim, C.D.Yoo and J.H.Lee, "Design of ultrasonic horn to enhance amplitude uniformity," *Journal of Korean Society of Precision Engineering* (2009), pp. 691-692.
5. Jeong-Seok Seo, Yoon-Jung Lee, Si-Young Beck and Dong-Sam Park, "Prediction of crack initiation and design of 40kHz blade horn for ultrasonic cutting," *Journal of the Korean Society of Manufacturing Technology Engineers,* Vol. 21 (2012), pp. 784-789.
6. S.R Kim, J.H Lee and Choong D. Yoo, "Design of cylinder horn for ultrasonic welding," *Journal of KWJS*, vol. 27, No.4 (2009), pp. 60-66.
7. K.Park, W.S.Jung and J.M.Ra, "Optimal design of a tool horn for micro–pattern forming using ultrasonic wave," *KSMTE Annual Spring Conference* (2012), pp. 47-47.
8. Jeong-Seok Seo, Yoon-Jung Lee, Jin-Wook Kim and Dong-Sam Park, "Design improvement and performance evaluation of 20kHz horn for ultrasonic welding," *Journal of Korean Society of Manufacturing Process Engineers,* vol.12, No.5 (2013), pp.135-140.

MATEC Web of Conferences **257**, 02010 (2019) https://doi.org/10.1051/matecconf/201925702010
PMAE 2018

Finite Element Simulation of Forming Process for Automotive Components from Advanced High Strength Steel

Yachita Peraprutchaya[1] and Surasak Suranantchai[1]

[1]*King Mongkut's University of Technology Thonburi, Department of Tool and Materials Engineering, 126 Pracha Uthit Rd., Bang Mod, Thung Kru, Bangkok 10140, Thailand*

Abstract. Advance high strength (AHS) steel has been widely used in the automotive industry and order to reduce the weight of automotive components that effect to reduce fuel consumption, but vehicles maintain safety. However, forming process with AHS steel has many problems which the main defect was spring-back. Therefore, the usage of finite element (FE) method simulation with kinematic hardening materials model sustains and improve production processes. Yoshida-Uemori (Y-U) model is the one of kinematic hardening materials model that show the great capability for prediction of elastic recovery behavior. Consequencely, this work has been research on AHS steel as the NSC980D grade for automotive components in the drawing process and applied FE simulation with Y-U model to improve the production process. Moreover, NSC980D steel was inspected in tension-compression and tensile tests were proceeded. As a result, material parameters from testing simulate with designed die and improvement were shown. Finally, comparing FE simulation forming with experiment forming, showed FE simulation with Y-U model is suitable for prediction and process improvement.

1 Introduction

Nowadays, the trend of automotive industry has been developed automotive components in terms of quality, cost and delivery for increasing competitive of manufacturer and responding the demand of customers. The consumer demand focus on fuel economy. Consequently, manufacturers try to use AHS steel materials into production process for the intention of reducing weight but safety remaining [1].

However, forming process of AHS steel materials has more problems in strength of materials. The necessary defect in drawing process is spring-back effect especially in high strength materials. It was found that the materials with higher strength is more spring-back than lower strength materials [2]. The solution of improvement in production process is rather difficult by which waste of time and cost. Hence, it is necessary to use FE simulation to construct kinematic hardening materials model for analysis and improvement of die design process. It becomes necessary for producing process to reduce time and cost in trial out and error of die design process [3].

Recently, it has been studying and creating a variety of kinematic hardening materials model to describe plastic deformation behavior and forming prediction of materials. Each materials model is suitable for different application. Y-U model is the one of kinematic hardening materials model that shown the great capability for description about elastic recovery behavior and spring-back effect prediction. Some researches for NSC980D steel grade and Panel-RF FRT HRD vehicle component that was formed by drawing process. Moreover, FE simulation with Y-U model has been used for improving spring-back effect. Including, tensile testing and tension-compression testing obtain materials parameters that required to simulation forming. For this research, the commercial program PAM-STAMP, based on FE, was applied to the die design. Finally, the result of FE forming simulation compare and analyze with experiment forming.

2 Materials model

This research concentrated to study on Y-U model with AHS steel NSC980D grade. Y-U model was constructed within two-surface plastic model of the Bauschinger effect, which is a kinematic hardening of yield surface and mixed isotropic-kinematic hardening of bounding surface. It has ability to describe the transient Bauchinger effect of high strength material deformation characterized by early re-yielding and smooth elastic-plastic transition with a change of work-hardening rate and the permanent softening characterized by stress offset in the region after the transient period, shown as Figure 1. And the work-hardening stagnation appearing at the range of reverse deformation of high strength material behavior [4]. For applying this model, it is required many parameters from the mechanical properties testing of NSC980D, which was seven plasticity materials parameters (Y, B, b, R_{sat}, C, k, and h) and three Young's modulus parameters (E_0, E_{sat}, and ξ). Y is the initial yield strength; B is the initial size of bounding surface, C is the material parameter for kinematic hardening rule of yield surface and its isotropic hardening component; R_{sat} is the saturated value of the isotropic hardening stress at infinitely large; k is a material parameter that control the rate of isotropic hardening; b is kinematic/isotropic hardening of bounding surface parameter; h is the materials parameter that controls the work-hardening stagnation; E_0 is the initial of Young's modulus; E_{sat} is the saturation value of Young's modulus when the strain becomes infinite; and ξ is the materials parameter that control the plastic-strain dependency of Young's modulus [5]. Those parameters of Y-U model can be obtained from the material testing data for AHS steel materials tests with tension-compression testing. In addition, anisotropy parameters will be determined by some of conventional tests as uniaxial tensile test, etc. [6].

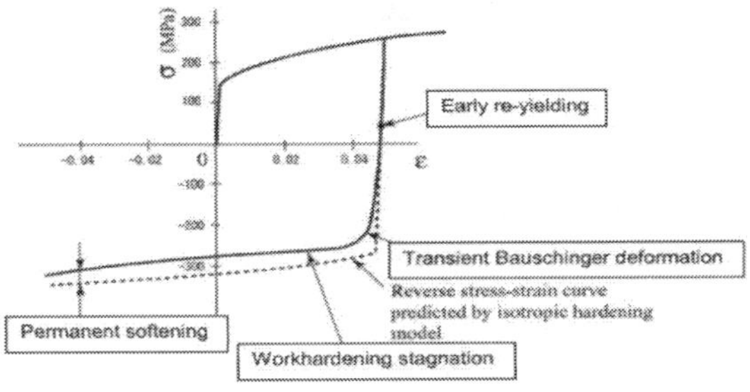

Figure 1. The illustration of two-surface plastic modeling [4]

3 Materials testing and experiment

In this research, the researchers tested on steel sheet grade NSC9890D, which is a dual phase steel, with 1.0 mm thickness. The mechanical properties of material were examined by tension-compression testing and uniaxial tensile testing for determine material parameters of Y-U model and some anisotropy parameters to apply with FE program.

3.1. Tension-compression testing

This testing was performed to total tension and compressive strain about 10 percent to obtain a saturated state of response to stress and strain. The tests describe Bauschinger effect during the reverse deformation, re-yielding, transient Bauschinger deformation, work-hardening stagnation and permanent softening behavior of AHS steel, that shown in Figure 1. The cyclic test has a spring fixture for prevent bending and has a digital image correlation (DIC) system for measuring the local displacement of specimen. From the cyclic test, the result is true stress – strain curve for determining of Y-U parameters, as concluding with Table 1. In addition, the material parameter that illustration the relationship of Young's modulus changing function and applied effective plastic strain, are given in Table 2.

Table 1. Material parameters of Y-U model for NSC980D steel testing from cyclic test

Y [MPa]	B [MPa]	b [MPa]	R_{sat} [MPa]	C	k	h
739.36	918.94	24.55	290.24	200	16.09	0.5

Table 2. The relationship between Young's modulus and plastic strain parameters [7]

E_0 [GPa]	E_{sat} [GPa]	ξ
189.91	128.40	14.19

3.2 Uniaxial tensile testing

Firstly, the uniaxial tensile testing was performed to determine the anisotropic plastic behaviors of steel examinations. The tensile testing was using hydraulic tensile test machine with constant crosshead speed of 0.05 mm/s and record the image during the test to estimate the plastic strain ratio (r-value) by DIC technique. In addition, the preparing of sheet specimens has three difference directions that were parallel, transverse, and 45° to the rolling direction. After the tests, the results determined the Young's modulus (E), 0.2% offset yield strength (Y), ultimate tensile strength (σ_u), total elongation (ε), and plastic strain ratio. Moreover, it has been able to determine the work-hardening exponent (n) and the strength coefficients (K) of parallel to the rolling direction, all parameters that shown in Table 3.

Table 3. Mechanical properties of NSC980D steel from tensile testing

Direction	E [GPa]	Y [MPa]	σ_u [MPa]	ε [%]	n	K [MPa]	r
0°	189.76	670.57	946.84	10.05	0.1024	1324.47	0.82
45°	188.99	761.03	1079.85	10.68	N/A	N/A	0.83
90°	190.98	770.18	1032.29	10.04	N/A	N/A	0.92
Avg.	189.91	733.92	1019.66	10.26	N/A	N/A	0.86

3.3 FE element

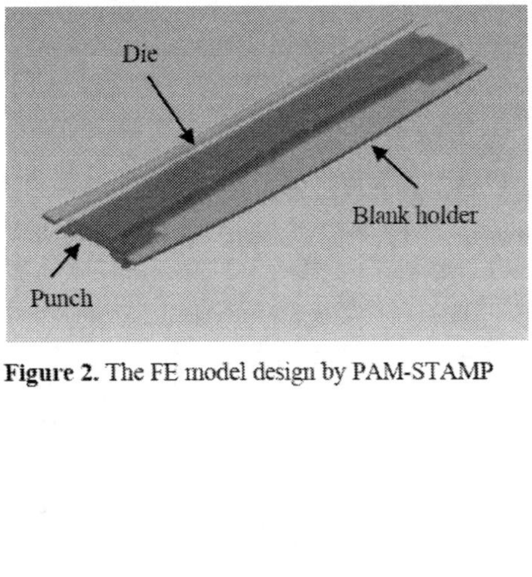

Figure 2. The FE model design by PAM-STAMP

In this work, study on Panel-RF FRT HRD component, it was automotive roof frame, and used the commercial program was PAM-STAMP to design 3D FE model, predicting and improving the die compensation for solve the spring-back effect. The position of tools shown as Figure 2. The thickness of blank was 1.0 mm by definite the blank size was 225 mm in width and 1000 mm in length. To create the finite element of the initial blank material with parameters of material model from mechanical testing, the mesh quality has been ensured to reliability for FE simulation. The initial mesh size was 14 mm and final mesh size was 0.875 mm. The refinement mesh size has maximum level was 5. The setting of drawing condition for FE simulation, has been set up with the stamping velocity was 5 m/s and the holding velocity was 2 m/s. The friction coefficient of die/punch and blank holder was definite to 0.1. The cushion stroke which the initial rise of the blank holder was set to 40 mm and while FE simulation forming that use blank holding force at 9 tons. The overall of drawing process is single action. And Figure 3 showed the FE simulation model step [5]. In Step one was the gravity step. The blank was placed on the blank holder by only gravity force. Next, the holding step was the upper die hold on the blank and contact the blank holder. Third, stamping step, the blank was deformed into final product of forming process. After the forming step, the spring-back simulation was proceeded. FE simulation in this research, has been used with material models for compare the spring-back prediction such as Y-U model, Hill'48 and 980_GEN3_GPa which was the materials parameters in PAM-STAMP

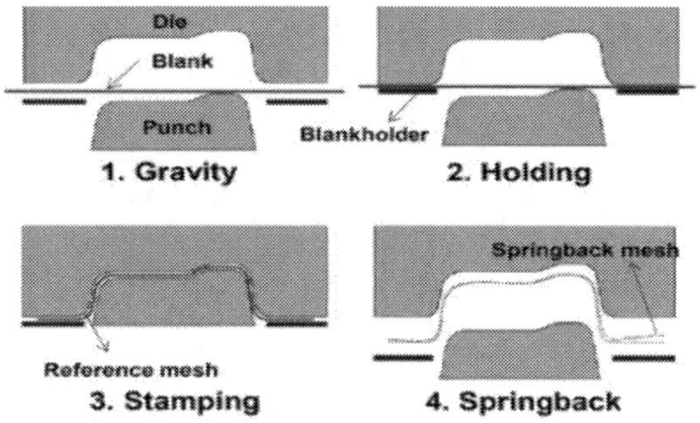

Figure 3. The diagram of forming state

3.4. Drawing process

To prove the material parameters from testing, Y-U model was predicted the spring-back effect. Automotive component was better than other model and close to the actual forming test with the same condition and die design as the forming simulation. The actual automotive component, Panel-RF FRT HRD is shown as Figure 4.

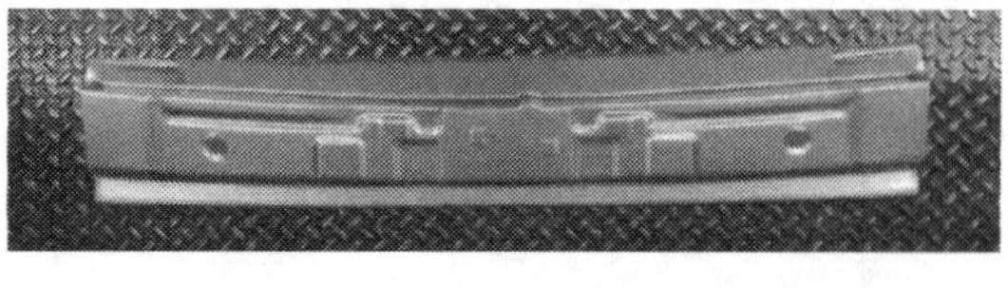

Figure 4. The actual product of Panel-RF FRT HRD

3.5. Spring-back comparison

In FE simulation and forming test of automotive component. The researcher used 3D file, which the actual product had been converted with 3D scanning, for compare the spring-back effect by cross-section of actual and simulation products. Considering with 3 sections, section A was 150 mm from the left, and section B was the center of product and section C was 300 mm from the right. Figure 5 showed each section that considered. The next step, the researcher used the sections to evaluation of angle variable to predict and compare the spring-back effect as shown in Figure 6.

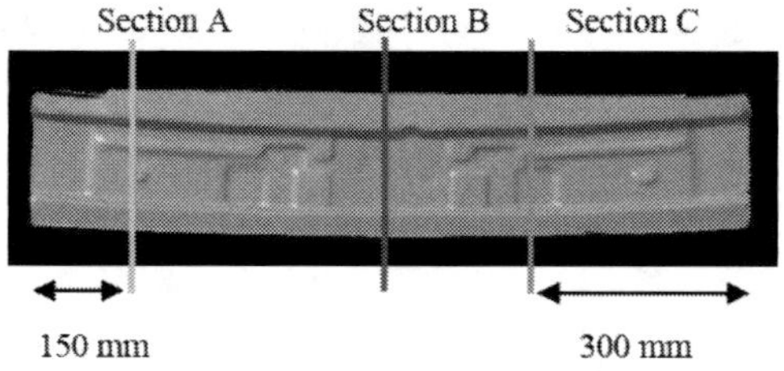

Figure 5. The section for spring-back comparison

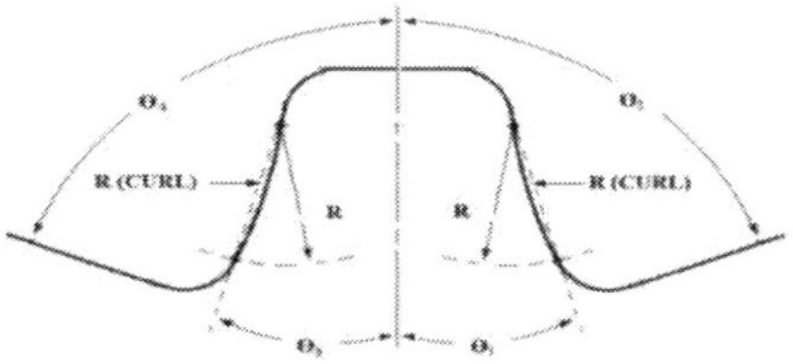

Figure 6. The variables for spring-back evaluating

4 Result and discussion

The research product performed by FE simulation with different material model such as Y-U model, Hill's 48 and 980_GEN3_GPa focused on spring-back effect considering with displacement along direction contour. A result of Y-U model simulation was observed that the local maximum and minimum deviation due to the spring-back were 6.844 mm and -5.239 mm by respectively at the end of product as shown in Figure 7.

Figure 7. The comparison of before and after spring-back effect of Y-U simulation

Then, Figure 8 showed the local maximum deviation of spring-back was 6.368 mm at the end of product and the minimum spring-back was –2.567 mm was more at the center of product that was the results of Hill's 48 model simulation.

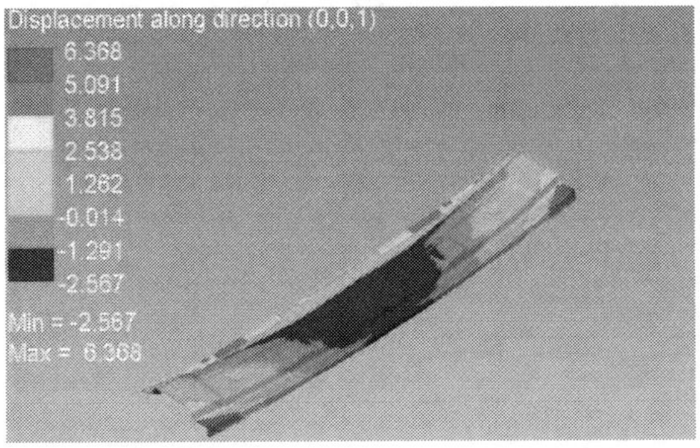

Figure 8. The displacement along direction of the Hill's 48 model simulation

For the 980_GEN3_GPa material parameters was detected that the local maximum and minimum of spring-back ordered was 5.368 mm and -1.784 mm, the area where has spring-back been similar with a result of Hill's 48 model. The displacement along direction of 980_GEN3_GPa simulation is shown as Figure 9.

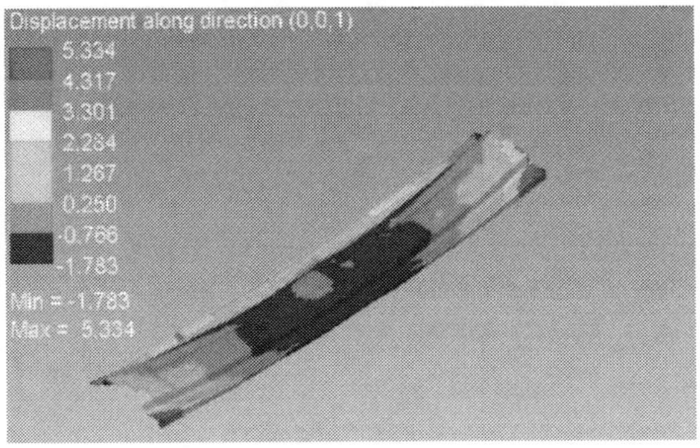

Figure 9. The displacement along direction of the 980_GEN_GPa simulation

In case of cross section profile for comparing the angle of spring-back from the experiment and simulation with each material model parameters showed that the result of Y-U material model simulation has been closely with actual experiment more than another, the Figure 10-12 shown as a comparison of experiment and each material model parameters cross section, in section A, B and C.

Figure 10. The comparison of cross section profile in section A

Figure 11. The comparison of cross section profile in section B

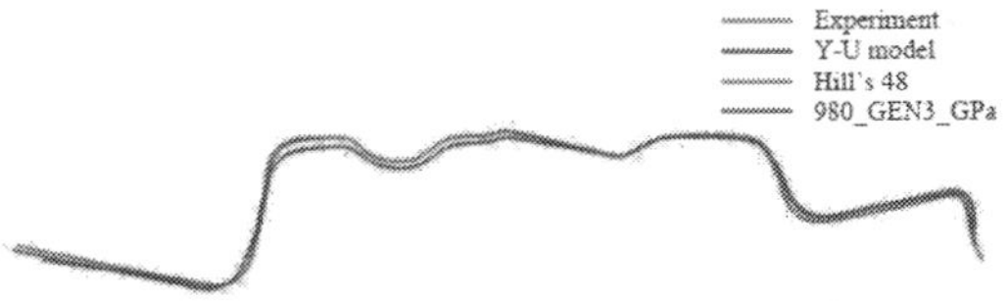

Figure 12. The comparison of cross section profile in section C

In Figure 13, it was clearly shown the difference in predictor of spring-back effect, when prepares with the actual experiment forming.

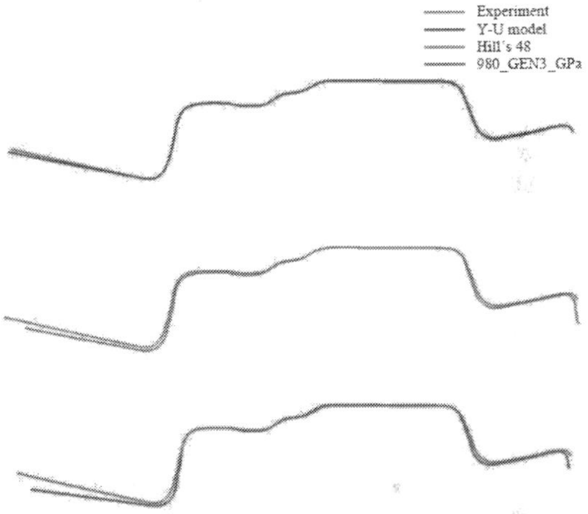

Figure 13. The comparison of each model section prepares with experiment in section C

And from the angle comparing, as shown in Figure 6, it is found Y-U model simulation has the effect of deviating from the actual part was less than another model; the maximum deviation was 4.84 degrees, the minimum deviation was 0.16 degrees and the average deviation was 1.26 degrees. While the Hill's 48 model has the maximum deviation was 5.56 degrees, the minimum deviation was 0.83 and the average deviation was 2.93 degrees. As same as the 980_GEN3_GPA has the maximum deviation was 5.35 degrees, the minimum deviation was 0.83 and the average deviation was 2.54 degrees.

5 Conclusion

From materials tested by static tensile and tension-compression made many materials parameters to use in materials model such as Y-U model and Hill's 48. Exceptionally, 980_GEN3_GPa was a material parameter from PAM-STAMP program. This research could apply to study on NSC980D steel for automotive component forming, which is Panel-RF FRT HRD. The investigation was found that Y-U model was appropriated to use finite element simulation for prediction the spring-back of AHS steel which was better than another materials model because research study was shown the result of Y-U simulation closely with actual experiment. This kinematic hardening materials model was consisted of seven plasticity material parameters and three parameters of Young's modulus parameters, which could be described the behavior of AHS steel.

6 Acknowledgment

This research has been funded by the Thailand Research Fund (TRF) and Research and Researcher for Industries (RRi). Moreover, the author was supported material from Thai Summit Automotive Co., Ltd.

References

1. W. Gan, S. S. Babu, N. Kapustka and H. Robert: Microstructural effects on the spring-back of advanced high-strength steel, Metallurgy and Materials Transations A. Vol. **37** (2006), p. 3221-3231.
2. P. Chen and M. Koç: Simulation of springback variation in forming of advanced high strength steels, J. Mater. Pro. Tech. Vol. **190** (2007), p. 189-198.
3. F. Pourboghrat and E. Chu: Prediction of spring-back and side wall curl in 2-D draw bending, J. Mater. Pro. Tech. Vol. **50** (1995), p.361-374.
4. F. Yoshida and T. Uemori, in:*Cyclic plasticity model for accurate simulation of spring-back of sheet metal*, 60 Excellent Inventions in metal Forming, edited by A. E. Tekkaya, W Homberg and A. Brosius Publising. Berlin, Heidlberg (2015). P. 61-66.
5. PAM-STAMP 2G Profession 64-Bit Program.
6. E. M. Lee, D. S. Shim and et al: Study on design of progressive dies for manufacture of aotomoblie strctural member using DP980 advancd high strengtg steel, J. Mech. Sci. Tech. Vol. **30** (2016), p. 853-864.
7. W. Julsri, S. Suranuntchai and V. Uthaisangsuk: Study of spring-back effect of AHS steel using a microstucture based modelling, Int. J. Mech. Sci. Vol. **135** (2017), p. 499-516.

AUTHOR INDEX